STOCHASTIC RESONANCE

Stochastic Resonance
Theory and Applications

Edited by

Bruno Andò
University of Catania,
Catania, Italy

and

Salvatore Graziani
University of Catania,
Catania, Italy

KLUWER ACADEMIC PUBLISHERS
BOSTON / DORDRECHT / LONDON

A C.I.P. Catalogue record for this book is available from the Library of Congress.

ISBN 0-7923-7779-6

Published by Kluwer Academic Publishers,
P.O. Box 17, 3300 AA Dordrecht, The Netherlands.

Sold and distributed in North, Central and South America
by Kluwer Academic Publishers,
101 Philip Drive, Norwell, MA 02061, U.S.A.

In all other countries, sold and distributed
by Kluwer Academic Publishers,
P.O. Box 322, 3300 AH Dordrecht, The Netherlands.

Printed on acid-free paper

Printed in the Netherlands.

To Our Parents

CONTENTS

FOREWORD

Noise is and has been the worst drawback in measurements, electronics, communications and control when we are dealing either with linear or non-linear systems.
The work of physicist and engineering is often devoted to design techniques and devices to avoid the undesired effects on noise.
Moreover dealing with particular non linear circuits the use of noise could be advantageous; this fact has been discovered twenty years ago. The term "Stochastic Resonance" was introduced to underline the characteristics of the systems where the outlined phenomena occur. The revolutionary concept is that noise could improve the performance of a system.
Added noise strategies, therefore, represent a new way to design non-linear circuits and systems with innovative suitable properties.
The book treats the noise-added systems in a complete and useful form. This new topic, extensively investigated in the monograph by the authors, is of particular interest in various areas of the science and of engineering.
Even if in the last decade a lot of papers have been written on the subject, the present work is the unique book that summarises the main results obtained in the specific field. It introduces new ideas to approach with the study of noise added systems, links the subject to the use of experimental chaos and presents in details very useful applications. It includes also a particular section devoted to the specific simulation tools developed by the authors that is very impressive.
The efforts of Dr. Andò and Dr. Graziani to organise this work has been, in my opinion, remarkable because it is not only a research manuscript but also a clear powerful text that could be appreciated by people interested in the real applications of the noise added systems from a practical point of view. The text could be considered an "Handbook of Noise Added Systems"!
Moreover, the main quality of the work is the clearness coupled to the scientific exactness.
The authors are very clear expositors both orally and in writing. It has been a pleasure for me to read their book.

Luigi Fortuna
DEES - Engineering Faculty
University of Catania, Italy
December, 1999

PREFACE

The book deals with the theory of *noise-added systems* and in particular with *Stochastic Resonance*, a quite novel theory that was introduced in the 80s to provide better understanding of some natural phenomena (e.g. ice age recurrence).
Following the very first works, a number of different applications to both natural and human-produced phenomena were proposed.
The book aims to improve the understanding of noise-based techniques and to focus on practical applications of this class of phenomena (an aspect that has been very poorly investigated up to now).
Based on the above-mentioned target the book is roughly divided into two parts. The First part deals with the essential theory of *noise-added systems* and in particular a new approach to *noise-added techniques* that allows to a number of strategies proposed in previous years to be unified. The proposed approach also allows *real-time control* of the noise characteristics, assuring optimal system performance.
In the Second part a large number of applications are described in detail, in the field of electric and electronic devices, with the aim of allowing readers to build their own experimental set.
The book comes with an educational software the authors have developed during their research activity in the field of *noise-added systems*. This reflects our hope to stimulate people to work in the field.
The book is organised as follows.

Chapter 1 gives an overview of the vast amount of literature that has been produced in almost 20 years in the field of *noise-added systems*, starting with the very first works on ice age recursion to arrive at ring laser and *SQIDD* systems. The aim of this chapter is to help the reader to find order in a mass of works that sometimes appear to deal with very different topics.

Chapter 2 deals with the theoretic fundamentals of *noise-added systems*. The chapter does not aim to develop one new theory. It is better give the information necessary for a clear understanding of the topics that follows.
In particular, the first part focuses on the theory of *Stochastic Resonance (SR)*, while the second part deals with *dithering*. Finally the authors would like to introduce their idea of the differences between *SR* and *dithering*, two techniques that are sometimes wrongly confused with one another.

Chapter 3 describes a totally new approach to *noise-added techniques* that combines a number of strategies proposed in previous years. The proposed approach allows to perform a *real-time control* of the noise characteristics, assuring optimal system performances, while meeting the system constraints.
The proposed approach has been applied to two different classes of systems. In particular, bistable systems have been considered, due to both their practical applications (e.g. conditioning blocks in measuring devices) and their importance in the first works on *noise-added systems*, which makes them benchmark systems.
In the following subsection quasi-linear systems are dealt with. It should be recalled, in fact, that all systems behave like *quasi-linear systems* if a narrow working range is considered.

Chapter 4 suggests a solution to the problem of noise generation. Production of the noise required to activate the system is, in fact, one of the main problems of *noise added techniques*. In this chapter the use of non-linear circuits is proposed to overcome this drawback.

Chapter 5 describes a number of practical applications dealing with both bistable and linear devices. All the examples considered have been chosen taking into account their usefulness in real systems. A number of the applications proposed can be directly exploited to improve measuring devices.

Chapter 6 describes the educational software the authors have developed in a number of years of research activity in the field of *noise-added systems*. The software itself is provided as part of the book.

Bruno Andò
Salvatore Graziani
December, 1999
Catania, Italy

ACKNOWLEDGEMENTS

We are proud to work with people that a few years ago were our teachers at this University. We wish to thank them not just because of the support they gave us during the preparation of this monograph, but also for the years they have spent to teach us the fundamentals of Science.
We wish to thank all the researchers of the Gruppo di Coordinamento "Misure Elettriche ed Elettroniche"- Sezione del GNRETE for all the opportunities they gave us to talk about noise added systems.
Also we are grateful to all the colleagues and friends of the Dipartimento Elettrico, Elettronico e Sistemistico of the University of Catania.
We thank all the researchers, a lot of them were students with us, that are now with ST-Microelectronics and that have developed with us some of the applications that are described in the book.
Last but not less important we appreciated the collaboration of a number of students that helped us with the realisation of the hardware devices and the software described in the book.

The Authors

PART 1 THEORY

1 REVIEW OF NOISE ADDED SYSTEMS LITERATURE

1.1 Introduction

When systems are forced by using any kind of input signal, a corresponding output signal that depends exclusively on the input and on the structural properties of the system is generally desired. Unfortunately the output of real systems depends on a number of undesired sources, both inherent and external to the system: the corresponding effect on the output signal is generally known as noise.
Noise comes in a number of different forms that influence both everyday life and laboratory experiments. Acoustic, thermal and electrical are but a few examples of the different kinds of noise that can be defined. Urban traffic is considered a cause of noise because it produces acoustic signals that interfere with quality of life. An interfering radio signal is noise because it can cover the news we are interested in. Transistors are affected by a noise signal that gives a bound to low limit resolution in measuring devices.
As the reported examples show, noise is generally not welcome because it limits system performance and great efforts, e.g. filtering, feedback compensation, opposing inputs, etc., are made to deal successfully with noise, even though at the expense of greater cost and complexity.
In the last decade a number of scientific studies have, surprisingly, shown the good side of noise. A number of natural phenomena can be explained if the presence of noise is taken into account (e.g. ice ages). In addition some physical systems work better in the presence of noise (e.g. biological neurons, electronic systems).
This surprising characteristic of Noise-Added systems and in particular Stochastic Resonance, a quite novel theory that was introduced in the 80s in order to gain better understanding of some natural phenomena, is the subject of this book.

1.2 Noise characterisation

In order to limit the negative effects of noise and to exploit its presence different approaches are adopted, based on the nature of the noise signal itself. As a matter of fact, though noise is largely unpredictable and not deterministic, some characteristics can be found that allow for a classification of different kinds of noise.

1.2.1 TAXONOMY OF NOISE

A well-known field where noise plays an important role is in electronics: readers who have been involved with the design of electronic circuits know well how much effort has to be devoted to reducing the undesired effects of noise. Even the simplest electric component, i.e. a resistor, produces a noise signal, if the real device is taken into account.

The noise that can be observed in the case of a resistor is due in large part to thermal fluctuations of electrons in the body of the device and is by far the commonest kind of noise that can be observed. This noise is called *thermal noise* or *Johnson noise* and is a form of *white noise* because, if its *Spectral Density* is considered, a flat spectrum is obtained.

A large set of phenomena, very different in nature, such as resistance fluctuations in semiconductors, the electrical activity of the heart, seismic activity, the level of insulin in diabetics, etc., produce a quite different noise signal, that is known as *flicker* or *pink noise* or *1/f noise.* The latter definition suggests that pink noise is predominantly a low frequency phenomenon. In Figure 1 examples of records of white *Gaussian*, white *uniform* and pink noise are given.

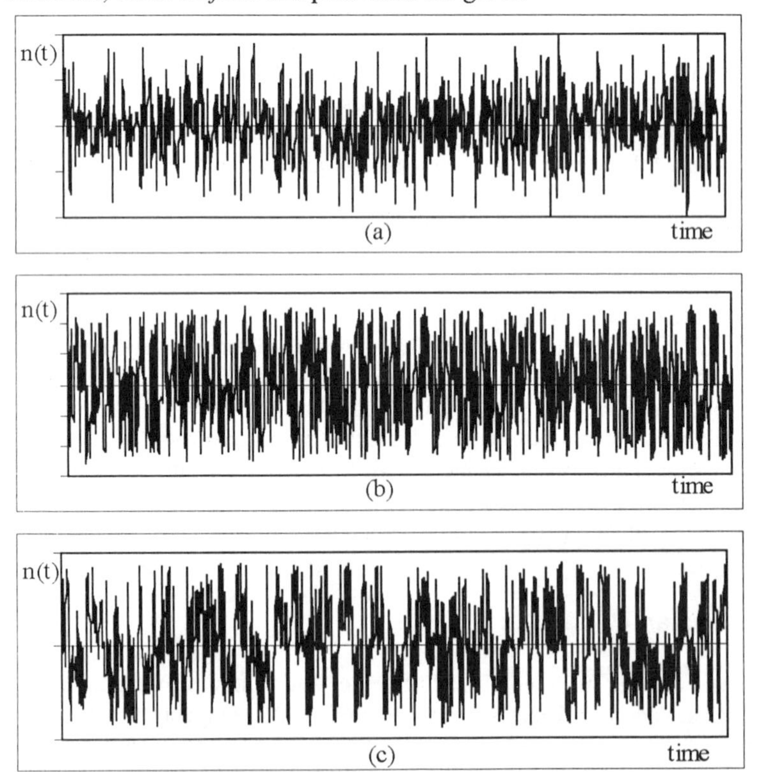

Figure 1. Records of white *Gaussian (a)*, white *uniform (b)* and pink noise *(c)*.

A number of other forms of noise have been introduced in different fields, such as *popcorn noise* and *Schottky noise* that affect linear integrated circuits (ICs).
The examples given above show that a taxonomy of noise can be obtained if its frequency content is considered. On the other hand, if the magnitude of the noise is taken into account a different classification is possible. It must be pointed out that the results obtained with the two approaches are not equivalent. As an example, if a white noise is considered and its *cumulative distribution* is computed, different results can be obtained, the most common cases being those of *Gaussian* and *uniform* distributions (this topic will be described in more detail in a following section).
Generally noise is considered a random signal, that is, a signal that cannot be described by a specific function ahead of its occurrence.
In the last few decades a number of works have shown that noise can also be the output of a chaotic system, i.e. of a deterministic system. This kind of noise has been called as *chaotic noise* and will be considered in the rest of the book. Figure 2 gives a schematic representation of the *Chua* circuit [1], its corresponding characteristic equation and its output signal. It is interesting to observe that the system output can be shown to be a noise signal.

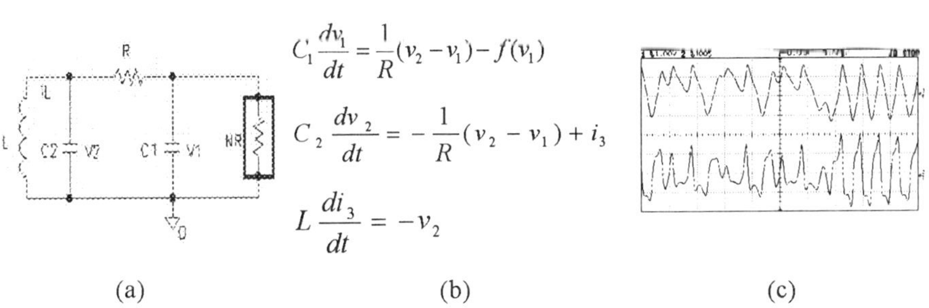

(a) (b) (c)

Figure 2. A schematic representation of a *chaotic system*: (a) the *Chua* circuit; (b) the system characteristic equation; (c) the system output signal.

1.2.2 MATHEMATICS OF NOISE

Figure 1 gives the time plot of three different kinds of noise (white *uniform*, white *Gaussian* and pink noise, respectively). It is very hard to establish the class of each signal, unless some kind of analysis is performed.
The random nature of noise (unless chaotic noise is considered) suggests that statistical analysis is the correct approach: it allows some statistical characteristics of the noise signal to be predicted. The most common methods involve either description of the signal magnitude or description of the rapidity of change of the signal.
Both approaches to noise description need some numerical parameters. In the rest of this sub-section some basic definitions will be given.

If we are given a random time variable *x(t)*, the *average* or *mean value*, is defined by

$$\overline{x(t)} \equiv \lim_{T \to \infty} \frac{1}{T} \int_0^T x(t)dt \tag{1.1}$$

Generally, we are concerned with a noise signal that has a zero mean value - if this is not the case, we can assume that the mean value defined by Equation (1.1) is subtracted from the total signal - and for this reason from here onwards the signal *x(t)* will be considered as having a zero mean value.
The mean value describes the constant component of the signal but does not give any indications of the signal magnitude: the *moments* of the random variable can be helpful. The *n*-th order central moment μ_n of the random variable is given by

$$\mu_n = \lim_{T \to \infty} \frac{1}{T} \int_0^T (x(t))^n \, dt \tag{1.2}$$

Of course, the first-order moment is the average of the variable.
The *second-order central moment*, also known as the *mean-squared value* of the signal, is by far the most widely used moment

$$\mu_2 \equiv \lim_{T \to \infty} \frac{1}{T} \int_0^T x(t)^2 \, dt \tag{1.3}$$

Also, to deal with a parameter that has the size of *x(t)*, the *root-mean-square* (*rms*), is generally used

$$x(t)_{rms} \equiv \sqrt{\mu_2} \tag{1.4}$$

A description of the way the signal values distribute around the mean value (whether it is zero or not) can be obtained by introducing the *Probability Density Function* $f(x)$. Definition of the *PDF* requires the introduction of the concept of probability, in particular the *relative frequency definition* (for different approaches to the definition of probability see [2]). For present purposes it will suffice to define the probability *P* that *x(t)* assumes values in a specific interval [x, x+Δx] by

$$P(x, x+\Delta x) \equiv \lim_{T \to \infty} \frac{\sum \Delta t}{T} \tag{1.5}$$

Δt being the total time spent by *x(t)* in the interval [x, x+Δx]. The P*robability Density Function f(x)* (*PDF*) is defined by

$$f(x) \equiv \lim_{\Delta x \to 0} \frac{P(x, x+\Delta x)}{\Delta x} \tag{1.6}$$

Though the *PDF* can take infinite different forms, provided it satisfies some mathematical constraints, some are more adequate to describe phenomena of real interest. The most widely-used *PDF* is the *Gaussian* or *normal* distribution, defined by

$$f(x) = \frac{1}{\sqrt{2\pi\sigma}} e^{\frac{x^2}{2\sigma^2}} \tag{1.7}$$

σ being the *standard deviation.* Its value depends on the corresponding second order moment. In particular, the following relationship holds

$$\mu_2 = \sigma^2 \tag{1.8}$$

Figure 3 shows how the difference between the *uniform* white noise and the *Gaussian* white noise shown in Figure 1 become dramatically apparent if the corresponding *PDFs* are considered.
Though the *uniform* noise has a totally different *PDF* from that of the *Gaussian noise*, this kind of analysis does not give any indication of the rapidity of change of the signals: two different signals can be given that show the same *PDFs* and have totally different time plots. An example of such a drawback is given in Figures 4 and 5. The signal shown in Figure 5a was obtained by grouping the data in Figure 4a into four groups and then ordering the data in each group in increasing order. It is therefore obvious that the two signals have the same *PDF*. The *PDFs* corresponding to the signal shown in Figures 4a and 5a are shown in Figures 4b and 5b.
The rapidity of change of random variables can be taken into account by introducing the A*utoCorrelation Function* (*ACF*) and the *power spectral density* (*PSD*) of the signal.

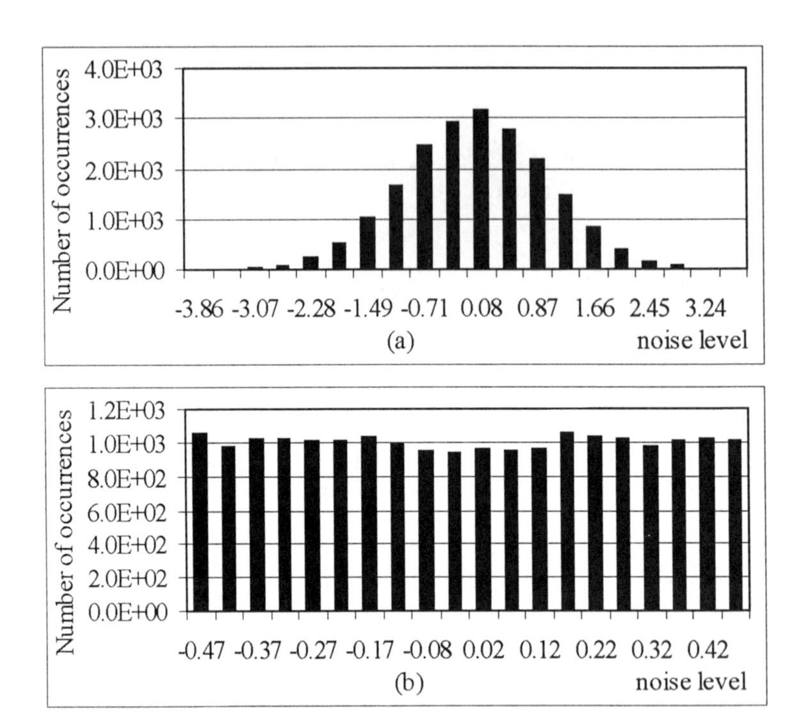

Figure 3. The Probability Density Function for a *Gaussian* (a) and a *uniform* (b) noise signal.

The *ACF* of the random variable *x(t)* is given by

$$R(\tau) \equiv \lim_{T \to \infty} \frac{1}{T} \int_0^T x(t)x(t+\tau)dt \tag{1.9}$$

The *ACFs* for the signals shown in Figures 4 and 5 are shown in Figure 6. It is evident that though both the *ACFs* show a large peak at $\tau=0$, variations in the value of τ produce small changes for the slow signal and very large changes in the case of the fast signal (a sharp maximum in the *ACF* is an indicator of a fast signal). Moreover it is worth observing that the *ACF* of a random signal at $\tau=0$ gives the corresponding mean-square value

$$R(0) = \mu_2 \tag{1.10}$$

The *PSD* conveys the same information as the *ACF* in a different domain, i.e. in the frequency domain. The *PSD* of a random variable *x(t)* is the Fourier transform of the corresponding *ACF*, as defined above

$$S(\omega)=\int_{-\infty}^{+\infty} R(\tau)e^{-j\omega\tau}\,d\tau \tag{1.11}$$

It can be shown that the *PSD* is a real function of ω and if *x(t)* is a real variable, as of course it is for the signals considered in this book, the *PSD* is real and even and can be computed as

$$S(\omega)=\int_{-\infty}^{+\infty} R(\tau)\cos\omega\tau\,d\tau=2\int_{0}^{+\infty} R(\tau)\cos\omega\tau\,d\tau \tag{1.12}$$

Of course, in some cases the inverse formulas are of interest and they can be found in any book on signal processing.

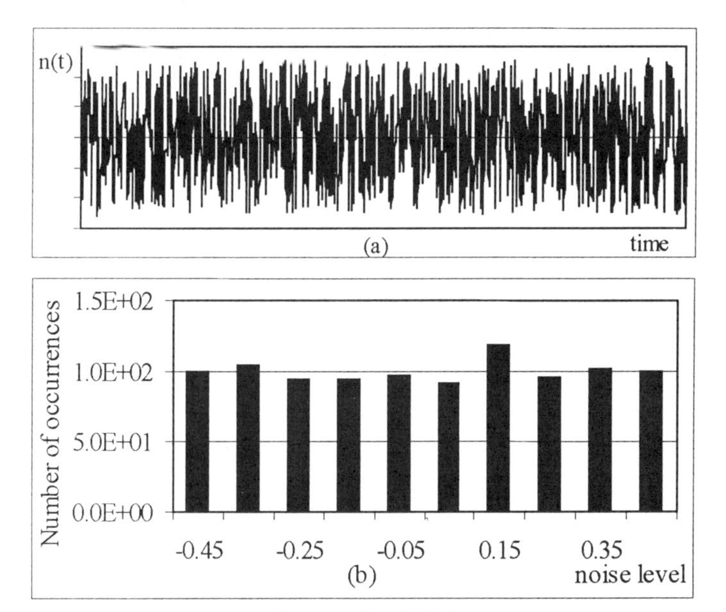

Figure 4. An example of white *uniform* noise (a) and its *PDF* (b).

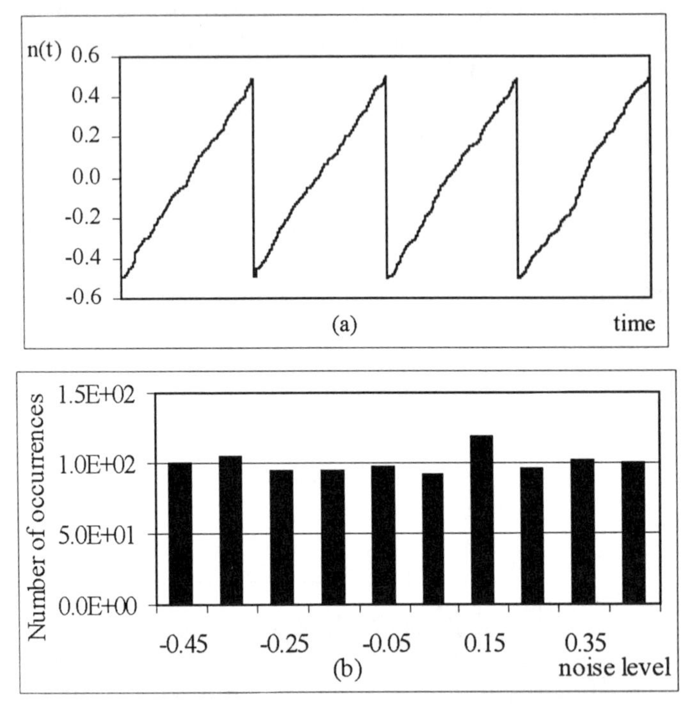

Figure 5. A graph of the data shown in Figure 4a, grouped into four groups and ordered in increasing order (a) and the corresponding *PDF* (b).

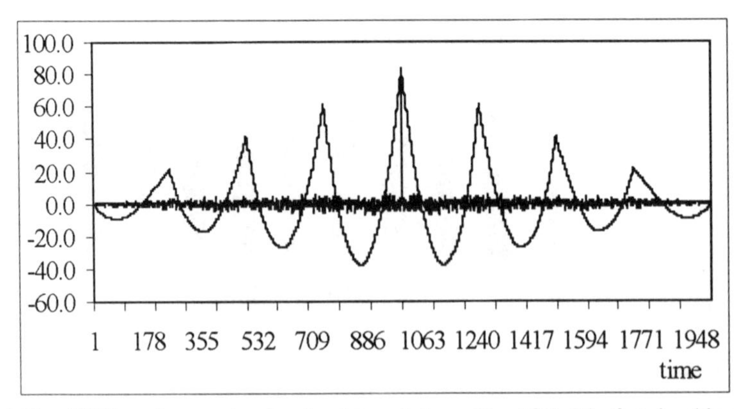

Figure 6. The *ACF* for a slow varying signal and for a fast one. The *ACF* of the fast signal has a sharper maximum at $\tau=0$.

A clear understanding of the meaning of the *PSD* can be obtained if the corresponding *Mean-square Spectral Density* is taken into account: the *PSD* of a signal is 2π times the corresponding mean-square spectral density, whose dimensions are those of $x(t)^2/(rad/s)$. Hence, if the mean-square spectral density is plotted versus the frequency ω, the corresponding area again gives the total mean-square value of the random variable (i.e. the power of the random variable). In the same way, if an interval of frequencies $\Delta\omega$ is considered the corresponding area gives the quantity of power lying within the considered frequency range $\Delta\omega$.
In Figure 7, the *PSDs* of a white noise and of a pink noise are shown. The differences between the two noises are now apparent as is the reason for their names.

1.2.3 SIGNAL RECOVERY FROM NOISE

A number of techniques have been designed to nullify or at least reduce the undesired effects of noise. In this section a brief description of the most usual methods is given [3].
The most widely-used method uses adequately selected filters to block spurious signals. The method can easily be understood if the spectra of both the desired and noise signals are taken into account. As shown in Figure 8 two different cases can occur depending on whether the spectra overlap or not.
The method can be successfully applied when the two spectra are widely separated. In this case the filter spectrum can be designed (i.e. the position of the zeros and/or poles of its transfer function) in such a way that the signal spectral components are left unaltered while the noise signal is blocked. An example of the results that can be obtained with the filtering method is given in Figure 9.
The figure reports the case of a noisy signal filtered by using a first order low-pass filter.
Unfortunately, the method does not work when the signal and noise spectra overlap either totally or partially. In this case, in fact, it is not possible to choose a filter 'window' that suppresses the noise and leaves the signal components unaltered. Moreover, the design of the filter can lead to high-order systems when the signal and noise spectra are very close to each other and/or high performance is required in terms of noise rejection, transaction band and signal distortion. In fact, the ideal filter should have an 'in-band' spectrum that does not modify the signal, its window shape should have a vertical slope and should totally block the noise out of the useful band. No real filter will have any of these ideal characteristics but generally the higher the filter order the better its performance.
The results that are obtained by using a second order low-pass filter, to eliminate the noise in the signal reported in Figure 9 are shown in Figure 10.
In cases where the filtering approach cannot be used, a number of different techniques can be helpful.
The best way to eliminate the influence of noise in any system is to design it in such a way that is inherently insensitive to undesirable inputs.

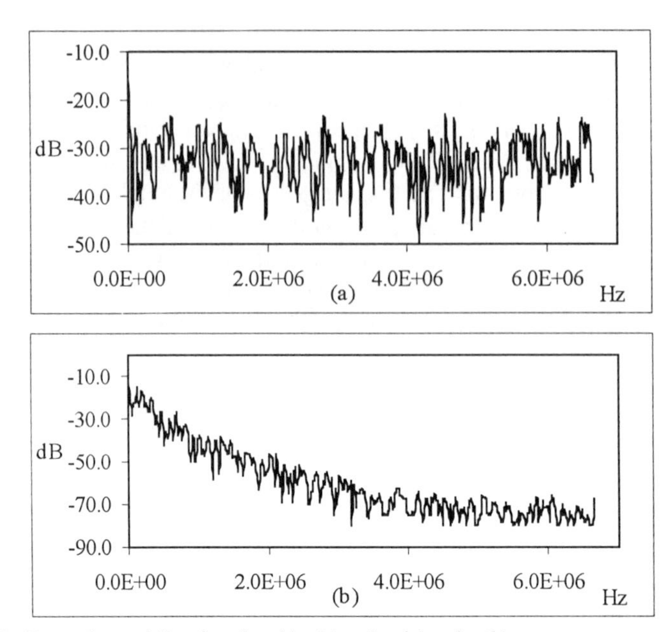

Figure 7. The Power Spectral Density of a white (a) and a pink noise (b).

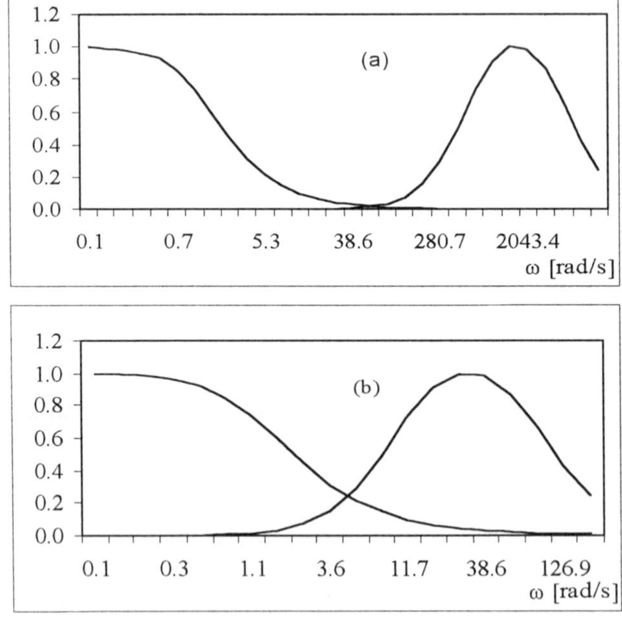

Figure 8. The spectrum plot of a signal and of noise: (a) the two spectra do not overlap; (b) the two spectra partially overlap.

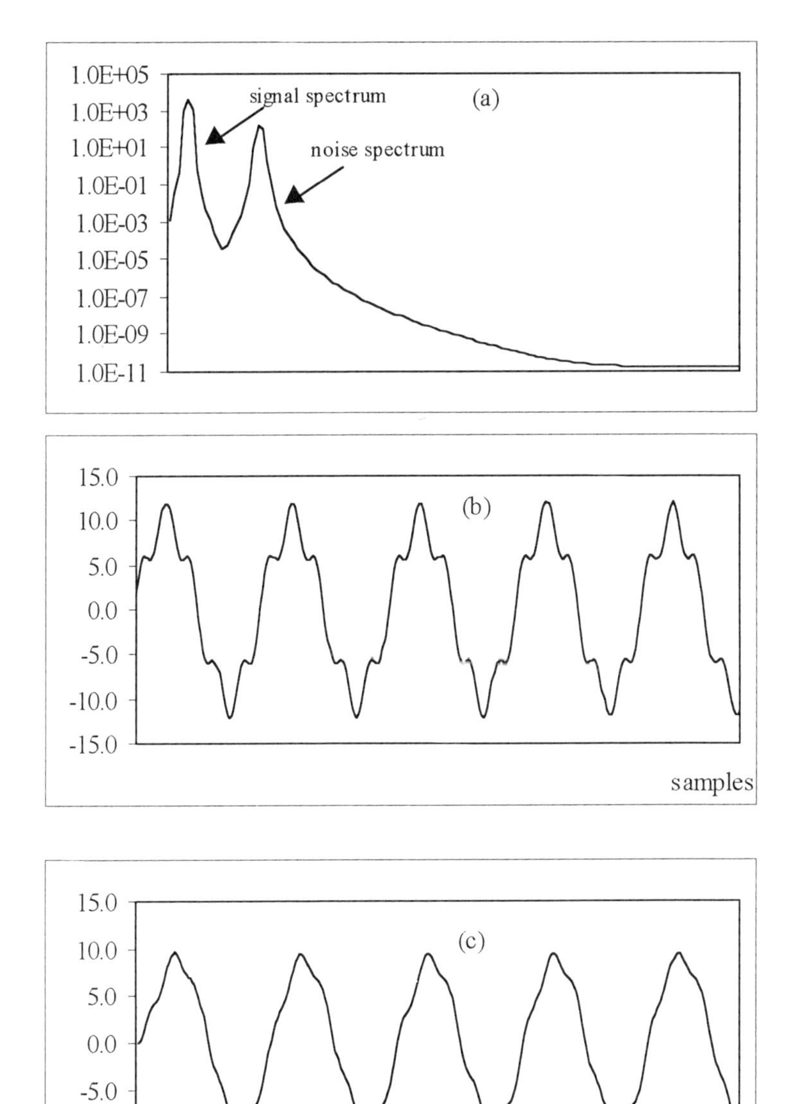

Figure 9. The *PSDs* of a noisy signal (a) and the corresponding signal filtered by using a first order *low-pass* filter: (b) unfiltered signal, (c) filtered signal.

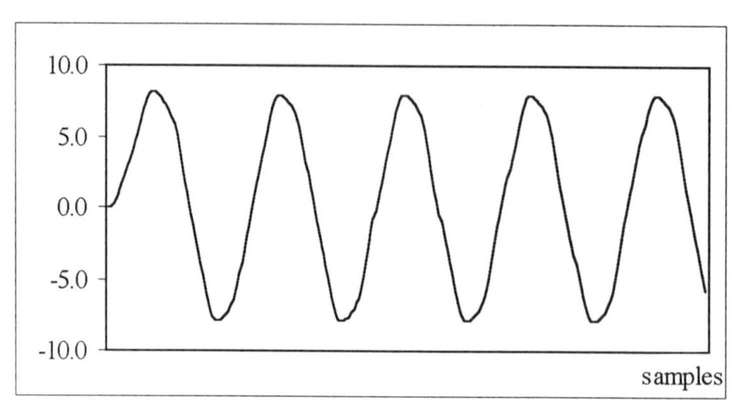

Figure 10. The noisy signal repoted in Figure 9 filtered by using a second order low-pass filter.

This approach generally requires a careful choice of the materials to be used and/or of the working principle.

Let us consider the case of a strain gauge that will be used to measure stress in a mechanical apparatus where temperature represents a undesirable input. In this case a material that has a very low thermal coefficient could greatly limit the effects of temperature fluctuations.

An application example that requires modifying the working principle is represented by systems that need to eliminate very slow varying signals (i.e. either bias or tilt due to temperature, time and line voltage). In this case an adequately chosen carrier signal is used to move the original signal spectrum in such a way as to allow the use of AC amplifiers, unaffected by either bias or drift. The resulting output is finally demodulated.

Figure 11 is a schematic representation of a strain gauge system that assures very low bias and drift. The system is forced by using a sinusoidal carrier signal and the unbalance signal of the bridge of strain gauges, the useful signal, is used to modulate the carrier. The system output is finally demodulated and low-pass filtered to obtain the desired information. Of course, in this kind of application the system bandwidth depends on the frequency of the carrier signal. The method of opposing and/or calculating input can also be used. In the first case a signal input that nullifies the effects of the undesired noise is intentionally applied to the system. In the second case the effects of the undesired noise on the output signal are estimated and subtracted from the system output.

Of course signal recovering is not the topic of this book. Readers who are interested in reading more about the techniques that have been developed to cope with this task can refer to [4].

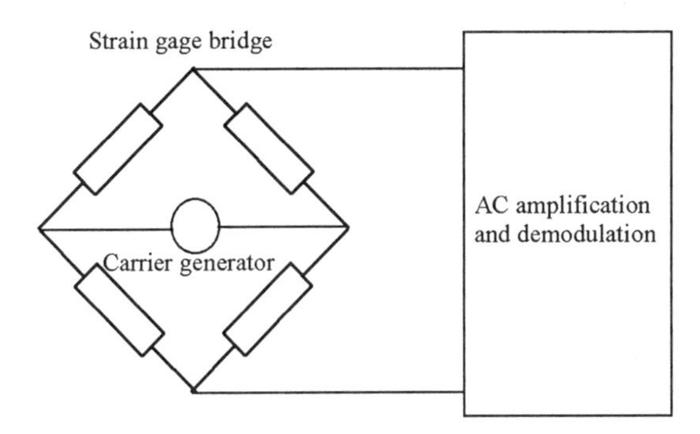

Figure 11. An example of amplitude modulation to filter out undesirable low-frequency noise.

1.3 Relevant bibliography on noise-added systems

In the last two decades the phenomena connected with noise-added systems have aroused the interests of a number of scientists, as confirmed by a bibliographical investigation of the topic.

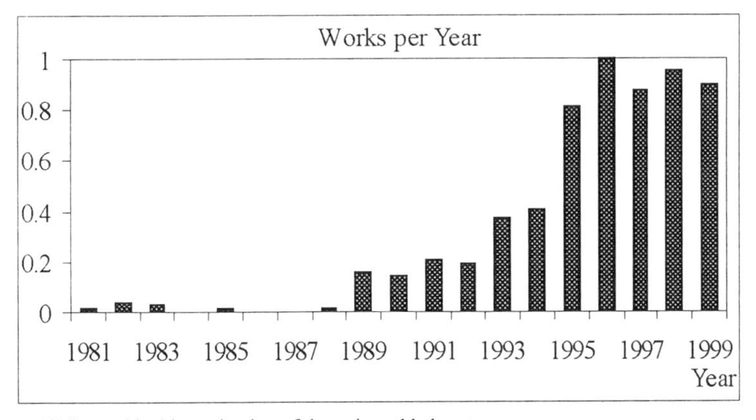

Figure 12. Bibliographical investigation of the noise-added system.

The results of this investigation are presented in Figure 12, which gives the distributions of scientific works on added-noise systems published each year from 1981-1999.

A list of the publications of most interest is given in Appendix A. Nearly 20 years of scientific investigation have been devoted to a topic previously considered to be

of marginal interest. Familiarity with the use of concepts linked to noise-added systems to interpret certain natural phenomena and the theoretical results obtained so far are essential for those who wish to work in this research field.
We therefore consider it useful to precede our discussion of noise-added systems with a description of the state of the art in this field. The references given in the Appendix have been subdivided into Introductory, Theoretical and Application studies.
For example, in *Benzi et al. 1981, 1982, 1983* and *C. Nicolis et al., 1981, 1982, 1983,* the authors propose an explanation of the phenomenon of the Ice Age using the concept of *SR*.

Benzi et al. 1981, 1982, 1983, C. Nicolis et al., 1981, 1982, 1983
The authors show that statistical analysis of variations in the volume of glaciers in the last 10^6 years has a characteristic variation period of 10^5 years. A possible natural cause that can be correlated with this event and has the same frequency is modulation of the eccentricity of the earth's orbit due to gravitational perturbation. The intensity of this phenomenon is, however, too low to be solely responsible for glacier formation. If, however, the weak exchanges of heat between the sun and the earth's surface are considered to be a further cause of the phenomenon, their interaction with modulation of the earth's orbit could explain the phenomenon. Benzi schematises this interaction by means of a potential with two wells in which one minimum represents the temperature of the surface covered with ice while the other represents the melting temperature. The slight modulation of the eccentricity of the earth's orbit is represented by a very low-frequency periodic component which makes the potential fluctuate, while the exchange of heat is represented by a noise with a *Gaussian* probability distribution. The interaction between these two signals enables the system to oscillate between one state and the other at the same frequency as the periodic signal, thus giving rise to the phenomenon of glacier formation.
The main feature of the studies classified as theoretical is that they deal with the phenomenon using mathematical and physical models of varying complexity, from the Langevin equation in *Benzi, 1981* to the unidimensional quartic potential model of Brownian movement in Marchesoni, Santucci et al., 1988, to a study of the dynamics deriving from use of noisy sources of various kinds in *Marchesoni et al., 1988* and *Gammaitoni, 1995.*
The aim of the following notes is to emphasise the importance of certain scientific papers published in past years for the topics dealt with in the rest of this book.

F. Marchesoni, E, Menichella-Saetta et al., 1988
This work illustrates the effect of coloured noise in Brownian motion.
The term Brownian motion is used to describe the movement of a particle in a liquid, subjected to collisions and other forces. Usually the collision force can be schematised by means of a white noise.

Observing the behaviour of this system forced by a white noise and a periodic signal it is possible to demonstrate the occurrence of *SR* phenomena.
The work proposes a general characterisation of the system by means of a spectral analysis. More specifically, it accounts for the presence of the two typical peaks in the spectrum -the *librational peak* and the *activation peak*- the effect of modulation of the *Arrenius factor*, equal to the relationship between the threshold and the noise amplitude, and the effect of variations in the friction factor. The remarks on the dependence of the escape rate on the noise correlation constant are particularly interesting and constructive.

L. Gammaitoni, F. Marchesoni et al., 1988
The paper is devoted to a description of the mechanism regulating the transitions of a stochastic system forced by a sub-threshold periodic component. More specifically, it describes the effect of oscillations of a quartic potential forced by a periodic component. On account of the symmetry of the system the two wells of the potential alternately take on two different levels. Two *Kramers Rates*, to which different meaning and significance can be attributed, are associated with the two levels.

M.I. Dykman, R. Mannella, 1990
The importance of this paper is more historical than scientific. The authors' remarks on *Gammaitoni and Marchesoni, 1989* are extremely interesting, and the interest aroused from now on in the phenomenon of *SR* is even more significant. *Dykman et al.* is an important example of this desire for knowledge and the synergistic efforts made to achieve this aim.

J.J. Collins, C. Carson et al., 1995
The authors demonstrate that a *FitzHugh-Nagumo* network of neurons has the capacity to detect a sub-threshold multi-frequency signal using a noise source with known characteristics. The result is really interesting because, in the system investigated, it separates the persistence of *SR* conditions from the noise amplitude. This is an extension of the aperiodic *SR* (*ASR*) phenomenon, observed in certain systems in the presence of aperiodic forcing signals, to complex systems.

S. M. Bezrukov, I. Vodyanoy, 1997
This is one of the first examples of extension of the concept of *SR* to non-dynamic systems with no threshold. The model the authors present is interesting on account of its simplicity and applicability to a number of physical and chemical systems such as p-n junctions.
For these systems, which are *normal*ly capable of processing signals with a low amplitude, the advantage of stochastic modulation lies in the possibility of enhancing the quality of the output signal. Performance enhancement is evaluated by means of the signal-to-noise ratio (*SNR*).

R. Dean Astumian, R.K. Adair, J. C. Weaver, 1997
This work is a speedy response to Bezrukov et al.. The criticism of the model of the non-dynamic system with no threshold is based on simulation. The *SNR*, calculated for a number of model parameter sets, does not seem to make the technique used particularly advantageous. *Bezrukov's* reply to this statement was quite decided but cautious: the system was still under investigation and so reservations were justifiable.

M. Dykman, R. Mannella, 1992
This work represents another significant contribution. By means of numerical calculation and experimental validation, the authors demonstrate the presence of a phase shift between the periodic forcing signal and the response of a stochastic system to which it is applied. They also show that the maximum phase shift is achieved in *SR* conditions.

The use of experimental validation obviously helps to make the concepts proposed less abstract. The first to present an experimental approach to the *SR* phenomenon, using a *Schmitt trigger, were S. Fauve and F. Heslot.*

S. Fauve e F. Heslot, 1983
Although the device used by the authors, a bistable comparator, is extremely simple to implement, it has a number of the characteristics typical of the phenomena being investigated.
Using an experimental approach, the authors introduce the concept *of "time matching*". They make an interesting and constructive criticism of the term Resonance as used in its most classical sense in deterministic systems.

D. F.Russel, L. A. Wilkens, F. Moss, 1999
The existence of Stochastic Resonance has been investigated also at the level of vital animal behaviour. In [5] the authors investigate, in fact, the possibility that paddlefish (*Polyodon spathula*) uses *SR* to locate and capture prey. In particular, they suggest that paddelfish uses passive electro-receptors to detect electrical signals from planktonic prey. Eventually the authors show that swarms of Daphnia plankton are a natural source of electrical noise.

As regards the application studies carried out by a number of researchers, we wish to stress the importance of their contributions to the development of stochastic systems. The opportunity on the one hand to validate the theories proposed, and on the other to confirm the existence of the phenomena being studied in real systems, provides those dealing with these phenomena with useful tools. The most significant contributions include the work on "ring lasers" by *McNamara et al., 1988*, the study on suppression of high frequency harmonics by *P. Jung and P- Talkner*, and

considerations on the persistence of *SR* conditions in systems processing signal "spikes" (e.g. *neural systems*).
Systems in which the application of a noise source leads to linearisation of their characteristics deserve a separate mention. They will be discussed in greater detail in later chapters, where it will be shown that these methods can be used to enhance the performance of measurement devices. The most representative work in this section includes *Carbone & Petri, 1994* and *Gammaitoni, 1995*.

References

[1] L. O. *Chua*, "Global unfolding of *Chua*'s circuit", IEICE Trans. Fundamentals, vol.E76-A, pp.704-734, May 1993.
[2] Papoulis, "*Probability, Random Variables, and Stochastic Process*", McGRAW-HILL BOOK COMPANY.
[3] Ernest O. Doebelin, "*Measurement Systems*", *McGRAW-HILL BOOK COMPANY*, third edition, 1985.
[4] T H. Wilmshurst, "*Signal recovery from noise in electronic instrumentation*", IOP, 1995.
[5] D. F.Russel, L. A. Wilkens, F. Moss., "Use of behavioural stochastic resonance by paddle fish for feeding", Nature, 402, Nov. 1999, 291-294.

2 AN OVERVIEW OF NOISE ADDED SYSTEMS

2.1 Introduction

In the last few years, the idea that extremely interesting phenomena can be generated by adding input noise to a non-linear system has found favour with a number of research groups [1-11]. Recent studies, for example, have focused on the behaviour of bistable threshold-based systems forced by both a periodic signal with an amplitude lower than the threshold and a stochastic component. The phenomenon involves a rapid increase in the signal-to-noise ratio, with an optimal noise variance value, and the presence of a peak in the output signal spectrum, corresponding to the frequency of the forcing signal. The phenomenon has been called *Stochastic Resonance* (*SR*) [1-21].

SR is completely different from the resonance introduced in linear systems. The most evident difference is represented by the value of the frequency characterising the resonance conditions. Whereas with *SR* the resonance frequency value is determined by the periodic signal, in the resonance typical of linear systems it depends on structural properties of the system. Later on we will discuss in greater detail the possibility of using the term *Resonance* for non-linear systems.

Having solved any confusion as to the meaning of the term, the possibility of enhancing the performance of a system by adding noise (e.g. stimulating a threshold-based system with a signal of an amplitude lower than the threshold, eliminating undesirable tones which may cause faults, enhancing the resolution of an A/D converter, etc.) comes as a pleasant surprise.

The persistence of the *SR* phenomenon in many physical systems existing in nature has been thoroughly investigated. Phenomena that classical theory was unable to explain have been interpreted by representing a complex system with a scheme having more simple characteristics in which co-operation between a deterministic and a stochastic source could account for its behaviour. Other research efforts have been devoted to force *SR* conditions to occur in systems of various kinds, so as to enhance their characteristics without intrusive action on the system.

As will be shown in the following sections of this chapter, the techniques to reduce the threshold zone in bistable systems can be used in applications where it is necessary to modify the threshold but not the topology of the system characteristics, for instance to output information concerning the harmonic content of a sub-threshold input signal (e.g. differential stage, comparators).

The first part of this chapter will deal with the problems involved in the *SR* phenomenon. Some applications of the relative techniques will be illustrated in Part II of the volume.
Another noise-added technique, normally used to enhance resolution in A/D converters, is *dithering* [22-28]. It is based on alteration of the quantisation error spectrum in such a way that its entity can be reduced by further processing. Although other techniques, such as oversampling, have been used to enhance the behaviour of these devices (e.g. to achieve a virtual increase in the number of bits in a A/D converter), dithering has proved to be the most effective tool in terms of the cost/performance ratio [25].
It should also be pointed out that if the concept is taken to extremes it is possible to interpret the increase in the resolution of an A/D converter as linearisation of its input-output features. This leads to the concept of linearisation of quantised systems with stochastic modulation techniques.
Let us consider a 1-bit converter. The output of the device can be at two levels. According to the amplitude of the input signal. If, for example, the amplitude is lower than the quantisation step the output will be a low-level one; if the amplitude is higher the output will also be high. Obviously such a system is insensitive to dynamic fluctuations in the input signal if they are lower than the quantisation step. Following application of a suitable noisy source, however, it is possible to enhance the sensitivity of the system to fluctuations in the dynamic of the input signal, even when the variations are lower than the quantisation step.
One possible application of these techniques is extension of the operating range of certain measuring devices (e.g. by reducing the threshold error or saturation of the system). More information on this topic will be given later on.
The general concepts of dithering will be dealt with in the second part of this chapter, while application of the techniques to measurement devices will be discussed in Part 2 of the volume.
From what has been said so far, it is clear that there are many advantages in stochastic modulation of a system in terms of performance improvement.
The attractiveness of noise-added systems must not, however, make us fail to consider the strange nature of stochastic sources. Careful assessment of the effects and advantages of these techniques will surely be of great help in understanding the phenomena typical of complex quasi-deterministic systems. It should not, in fact, be forgotten that by stimulating a system with a stochastic source the aleatory content of the forcing signal is inevitably transferred to the output, causing undesirable phenomena that often have to be foreseen and avoided during the design stage.
In applying noise-added techniques, therefore, it is necessary to respect the constraints imposed by the class of systems involved. Bistable systems, for example, which are normally used to study the phenomena involved in the *SR* process, given

that they only switch between two stable states, are capable of attenuating stochastic oscillations that respond to certain amplitude constraints. In this kind of system, therefore, performance enhancement is constrained by the limits imposed on the amplitude range of the stochastic signal.
The situation is quite different with pseudo-linear systems, which feature inherent non-linearity, where the noise fluctuation can be easily transferred to the output. The possibility of improving performance in these systems depends not only on the characteristics of the noise source but also on the use of devices to eliminate the stochastic component of the output.
It is therefore of fundamental importance to develop techniques to optimise the parameters of the stochastic component to be forced into the system.
In Chapter 3, to stress the importance of this statement and propose an alternative to the techniques described in the literature, we will present a strategy of general validity to determine optimal operating conditions for a generic stochastic system.
A final note concerns the possibility of applying the above-mentioned techniques to optimise the behaviour of systems operating in noisy environments. The basic idea is to use the environmental noise to improve device performance. Here it is no longer a case of optimising the characteristics of the noise, but of adapting and designing the system in such a way that performance can be improved in the presence of noisy sources with well-defined characteristics. Some consideration about this topic will be given later on.

2.2 The Stochastic Resonance Theory

Stochastic resonance is a phenomenon typical of certain non-linear systems, especially threshold-based systems.
Threshold-based systems feature one or two stable states in which the system remains for an indefinite period of time, unless it receives an external stimulus, the entity of which depends on the value taken by the parameters characterising the system. The passage from one state to the other is constrained by the input signal's capacity to supply enough energy to overcome the potential barrier.
In the last few decades numerous research groups have focused on the attempt to outline the theory behind the behaviour of systems forced by stochastic and deterministic signals, in particular systems with two or more stable states. *SR* defines a particular condition in which the spectrum of the output signal of a bistable system subjected to a periodic and a stochastic signal, both with an amplitude lower than the threshold, presents a peak in the proximity of the frequency of the periodic signal. This shows that the information associated with the harmonic content of the input signal has been transferred to the output.

an optimal forcing signal frequency value that will maximise the probability of the system synchronising with the forcing signal [26].
If a bistable system is forced by a periodic signal and a noise, the waiting time between two transactions can be monitored. The distribution of this quantity shows several peaks decreasing in amplitude. The first peak corresponds to the forcing frequency ν while the other peaks are odd multiples of 2ν. The area E under the first peak corresponds to the probability that the system is synchronised with the periodic signal. Of course the value of E changes according to the system parameter considered. For example, if noise variance is considered, E will reach a maximum value with $\sigma=\sigma_{opt.}$ In particular it is interesting to observe that if the quantity E is plotted as a function of the forcing frequency ν it shows typical resonant behaviour, thus confirming that when all the stochastic system parameters (noise variance, forcing amplitude, structural parameters) are fixed, a resonant frequency can be defined [26].
One of the first contributions towards an understanding of *SR* phenomena dates back to 1981, when the attempt was made to account for the phenomenon of glaciation [2].
As illustrated in Chapter 1, a number of further contributions have been made over the years, giving an increasing number of concrete examples in which the *SR* phenomenon plays an important role. Various subsequent theoretical studies have given us greater understanding and at times an analytical explanation of such phenomena.
Today it would be really surprising to find a system in which the SR phenomenon does not occur.

2.2.1 STOCHASTIC RESONANCE IN BROWNIAN MOTION

The Brownian motion model is used in many branches of physics as a prototype for the comprehension of various natural phenomena [8-9]. Several studies have emphasised the features of this phenomenon and its complexity due to its sensitivity to structural parameters [9]. The term Brownian motion is used to describe the movement of a particle in a fluid, subjected to collisions and other forces. Macroscopically, the position $x(t)$ of the particle can be modelled as a stochastic process satisfying a second-order differential equation

$$m\,\ddot{x}(t) + \gamma\,\dot{x}(t) + \dot{V}(x) = e(t) + f(t) \tag{2.1}$$

where m is the mass of the particle, γ is the coefficient of friction, $V(x)$ is a general potential function, $e(t)$ is the collision force, and $f(t)$ is a generic forcing term. On a macroscopic scale, the process $e(t)$ can be viewed as normal white noise with zero mean and power spectrum $S(\omega)=2kT\gamma$, where k is the Boltzman constant and T is the absolute temperature of the medium [29].

where m is the mass of the particle, γ is the coefficient of friction, $V(x)$ is a general potential function, $e(t)$ is the collision force, and $f(t)$ is a generic forcing term. On a macroscopic scale, the process $e(t)$ can be viewed as normal white noise with zero mean and power spectrum $S(\omega)=2kT\gamma$ where k is the Boltzman constant and T is the absolute temperature of the medium [29].
Very interesting behaviour of the system (2.1) has been observed when $V(x)$ is a Quartic Double Well potential [8, 9]

$$V(x) = -a\frac{x^2}{2} + b\frac{x^4}{2}, \text{ with } a > 0 \tag{2.2}$$

The minima of $V(x)$ are located in $x_+=(a/b)^{1/2}$ and $x_-=-(a/b)^{1/2}$. These are separated by a potential barrier with amplitude $DV=a^2/(4b)$. The characteristic frequencies of the system are: $\omega_b=[V''(x_\pm)]^{1/2}=(2a)^{1/2}$ and $\omega_0= [V''(x_0)]^{1/2}=(a)^{1/2}$. Figure 1a shows the shape of $V(x)$.
The system described exhibits the Stochastic Resonance phenomenon that in this system can easily be explained. If $f(t)=0$ the particle is subject to fluctuation forces $e(t)$. This force causes transitions between the two potential wells with a rate given by the *Kramers Rate*

$$R_K = \frac{\omega_0\omega_b}{2\pi\gamma}e^{-\frac{DV}{D}} \tag{2.3}$$

$D=KT$ being correlated to the noise fluctuation amplitude.

To provide a qualitative explanation of the dependence of the R_K on the noise characteristics, Figure 2 shows a typical output of a bistable system forced with a *Gaussian* white-spectrum stochastic component.
By varying the standard deviation σ it is possible control the amplitude of the noise fluctuations. Let us also assume that the value of σ is such that it allows the system to switch from one state to the other. Obviously these transitions will form a stochastic sequence and so it will not be possible to recognise a deterministic component in it. This behaviour, as shown in Figure 2, is accounted for by the fact that the instants at which the amplitude of the input signal is such as to overcome the threshold are totally random and so uncorrelated. It is, however, possible to determine a *Kramers Rate* or average switching frequency.
Determination of the analytical expression of the R_K is generally not a simple operation, but its dependence on the characteristic parameters of the system can be obtained by means of probabilistic considerations.

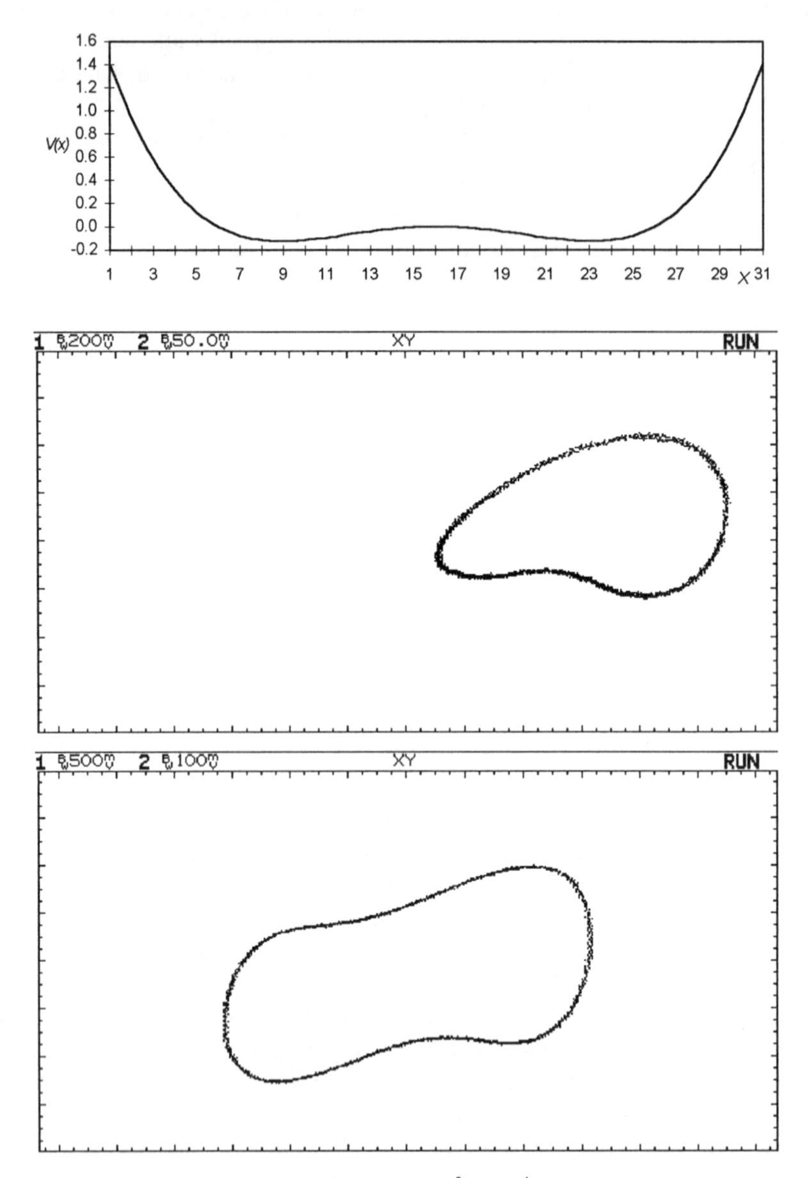

Figure 1. (a) The Quartic Double Well: $V(x) = -a\frac{x^2}{2} + b\frac{x^4}{2}$; (b) the state diagram when a sub-threshold forcing signal is applied; (c) the state diagram when a suitable forcing signal larger than the threshold is applied.

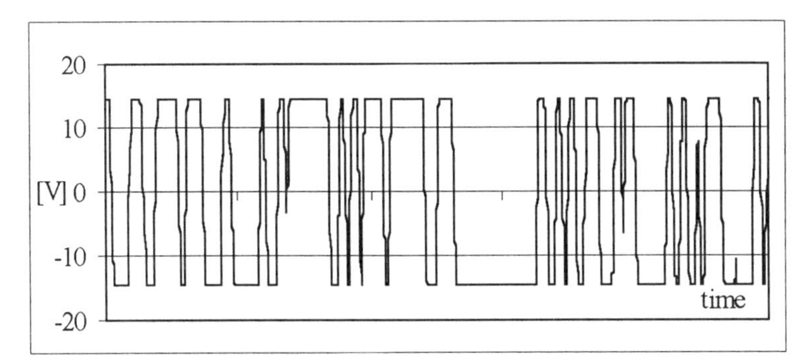

Figure 2. Output of a bistable system forced by a stochastic source.

Let us assume that we are sampling the output signal of a bistable system that switches its state due to added noise. A typical spectrum for this output signal is shown in Figure 3a. As can be seen, the spectrum is *lorentzian* [12-20] and can be represented with a law of the following kind

$$S(\omega)=f(1/\omega^2) \tag{2.4}$$

In this spectrum it is possible to identify a frequency R_K (*Kramers Rate*) corresponding to which the spectrum takes a value equal to half the maximum value.

Repeating the same experiment with varying stochastic signals, it is possible to find the following relation of proportionality between R_K and σ

$$\log R_K \propto -\sigma^{-2} \tag{2.5}$$

Hence

$$R_K \propto e^{-1/\sigma^2} \tag{2.6}$$

This relation is of great importance as it gives a formal basis for the possibility of controlling the average rate of stochastic transitions of a bistable system by varying the energy of the noise. Unfortunately, however, only in a few cases is it possible to make expression (2.6) explicit although it is theoretically valid, it is often of little practical use.

In the presence of a strong periodic forcing signal $f(t)$, with frequency $\nu=1/T$, the potential is tilted back and forth, alternately raising and lowering the potential barrier of the right and the left well. As a consequence, the particle rolls periodically from one potential well into the other one with the same frequency as

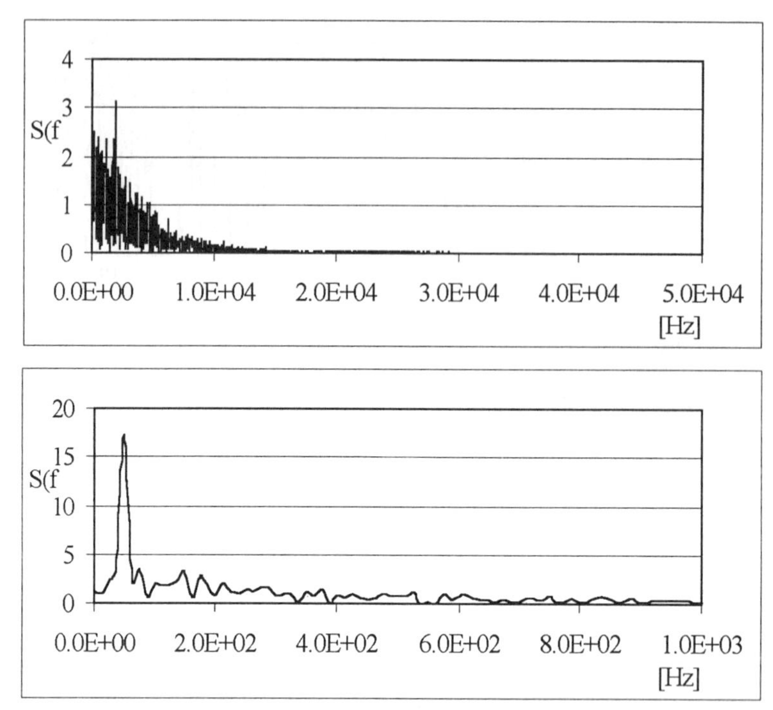

Figure 3. Spectrum of output signal of a bistable system forced by a stochastic source/and a sub-threshold periodic signal.

the forcing signal. The state behaviour of the system for this case is shown in Figure 1c.

In the presence of a weak periodic signal the particle should lie in a narrow range (attractor) of one state (which is selected by the initial conditions), as illustrated in Figure 1b. The presence of a suitable noise level signal could allow the particle roll from one potential well into the other one, this transaction being synchronised with the periodic forcing: the *SR* phenomenon. This statistical synchronisation takes place when the average waiting time between two noise-induced inter-well transitions ($T_K=1/R_K$) is comparable with half the period of the forcing signal T

$$2T_K(D)=T \tag{2.7}$$

The phenomenon can be interpreted by observing that the trend of the two thresholds of the bistable system will follow the periodic signal, with an average value of *S1* and *S2*, respectively, as shown in Figure 4.

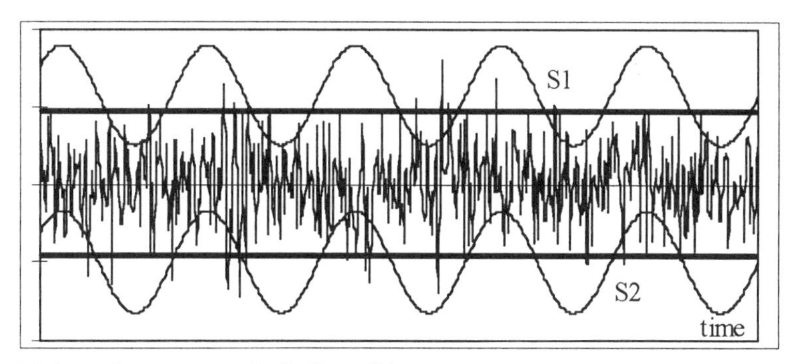

Figure 4. Scheme of a system operating in *SR* conditions.

Thanks to the co-operation between the two forcing signals, the system threshold is overcome a certain number of times in each semi-period. It is therefore becoming clear that although the system's behaviour is influenced by the stochastic signal, in this case it is closely correlated with the spectral content of the periodic forcing signal and its output follows a periodic trend, with a period of $T=1/\nu$.

When co-operation between a periodic and a stochastic signal provides the system with the periodic dynamic, the system is said to be operating in Stochastic Resonance conditions.

2.2.2 THE RESIDENCE TIME

The power spectrum of a bistable system's output signal, when forced by a noisy component alone, follows a trend similar to a *lorentzian* spectrum, with a law like $S(\omega)=f(1/\omega^2)$, where $S(\omega)$ represents the spectrum, $f(\cdot)$ is an appropriate function and ω the frequency. Spectrum of the output signal when the system is forced by a sub-threshold periodic signal modulated by a stochastic component presents a peak in the power spectrum corresponding to the frequency of the periodic signal ν. The spectrum of the output signal is shown in Figure 3b.

An equivalent way to characterise the behaviour of a stochastic system subjected to a periodic forcing signal is based on conversion of the continuous stochastic process $x(t)$, corresponding to the system output, into a punctual stochastic process, t_i. The instants t_i are those at which the output switches from one state to the other, i.e. the "*first crossing time*". At this point it is possible to define the quantity $T(i) = t_i - t_{i-1}$. The interval T represents the residence time between two subsequent transitions. In symmetrical bistable systems forced by stochastic signals alone, the distribution of $T(i)$ is a Poisson exponential distribution of the following form

$$N(T) = \frac{1}{T_K} e^{-\frac{T}{2T_K}} \tag{2.8}$$

Figure 5 shows this distribution and the information given is similar to what can be deduced from the spectrum shown in Figure 3a.
In the case of systems forced by both stochastic and periodic signals, *N(T)* presents peaks corresponding to the period of the forcing signal.

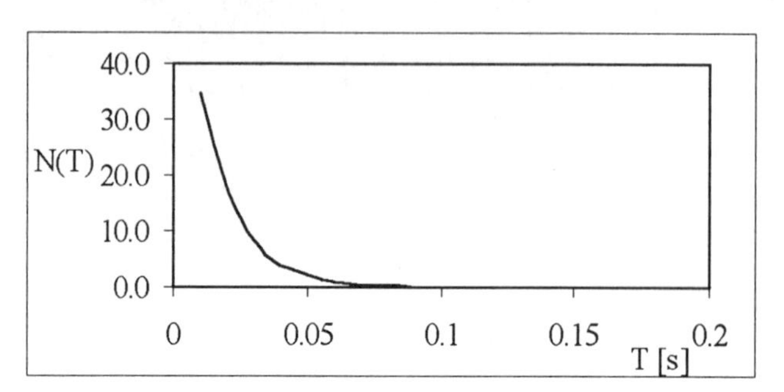

Figure 5. Distribution of residence time in a symmetrical bistable system forced by a stochastic signal.

2.2.3 THE SIGNAL-TO-NOISE RATIO

In a noise-added system it is convenient to define the signal-to-noise ratio as the follows

$$R=S_p(\nu)/S_n(\nu) \tag{2.9}$$

where
$S_p(\nu)$ indicates the peak of the output signal spectrum, calculated at the frequency, ν, of the periodic forcing signal;
$S_n(\nu)$ indicates the output signal spectrum at the same frequency, ν, but without the periodic forcing signal.

Figure 6 shows a typical trend for R plotted against σ^2/S, where S represents the system threshold. Analysis of the figure shows that corresponding to a particular σ^2/S ratio value, i.e. σ^2_{opt}/S, the ratio R presents a maximum zone.

The quantity σ^2_{opt} represents the noise variance value that allows the system to switch periodically without being affected by the stochastic trend of the noise. This value is also independent of the values taken by A. It is important to point out that σ^2_{opt} can be obtained experimentally by determining the values taken by R as σ varies.

The presence of a peak in the output signal spectrum and a maximum zone in the signal-to-noise ratio fully characterises the concept of Stochastic Resonance.

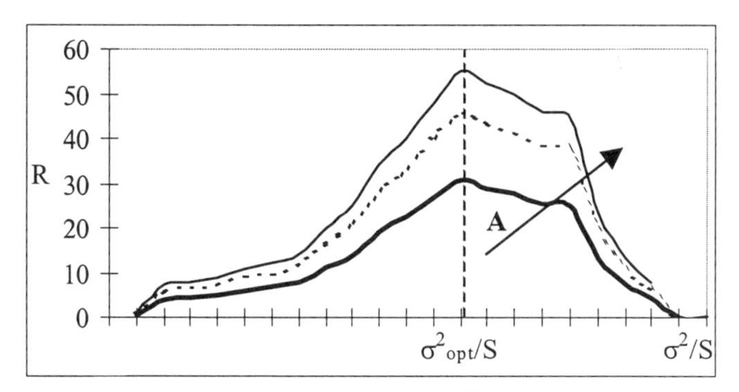

Figure 6. The signal to noise ratio as a function of the σ^2/S quantity, for different forcing signal amplitude.

2.2.4 THE INFLUENCE OF THE FORCING SIGNAL PARAMETERS ON STOCHASTIC RESONANCE CONDITIONS

In the following paragraphs we will deal with the effects of variations in the (periodic and stochastic) forcing signals on the persistence of *SR* conditions.

- *Forcing signal frequency: justification of the constraint* $\nu << R_k$.

In the literature it is often stated that for the *SR* condition to exist it is necessary for the noise bandwidth to be much wider than that of the deterministic forcing signal. In this case, in fact, with reference to what was stated previously, in each semi-period $T/2$ the combination of the two signals (at least a certain number of times) will be such as to guarantee that the threshold is exceeded. This condition is true for the example illustrated in Figure 4. Figure 7, on the other hand, illustrates a case in which, as the noise bandwidth is extremely limited, the probability of the occurrence of a transition synchronous with the periodic signal is much lower.

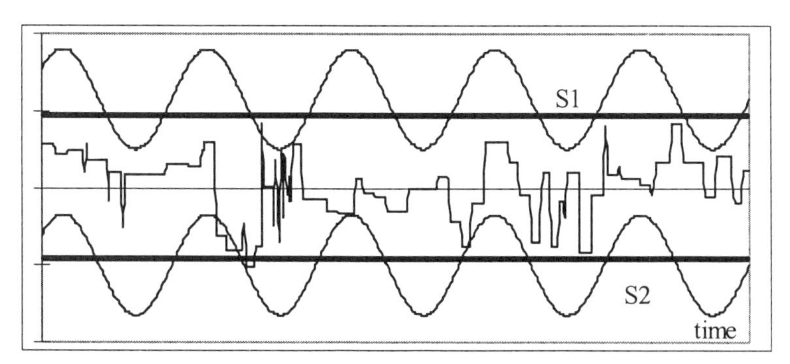

Figure 7. Interaction between the periodic signal modulating the threshold and the stochastic signal at a frequency $\nu \approx R_K$.

- *The optimal noise standard deviation value*

With the same threshold S, the peak in the ratio R always corresponds to a certain value $\sigma^2 = \sigma^2_{opt}$, irrespective of the value of A.

This can be accounted for as follows:

when $\sigma^2 > \sigma^2_{opt}$ the amplitudes of the stochastic signal oscillations become predominant and consequently the transitions do not only occur when the interaction between the two signals is such as to exceed the threshold, but in a random sequence;

when $\sigma^2 < \sigma^2_{opt}$ the opposite phenomenon occurs, i.e. the stochastic signal oscillations no longer trigger off transitions.

The amplitude of the forcing signal plays a fundamental role in the occurrence of the resonance condition, as will be illustrated in the following section.

- *The amplitude of the forcing signal*

Analysis of Figure 6 shows that as A grows, the value of the resonance peak also increases. With low A values, on the other hand, the resonance condition disappears. This is linked to the capacity of the forcing signal to interact with the noise oscillations and places a lower bound on the value of A which will allow an *SR* regime to be set up in the system.

Therefore, to operate in resonance conditions, the amplitude of the forcing signal has to take at least a certain minimum value, A_{min}.

At values over A_{min}, the *periodicity condition* improves.

The term *periodicity condition* means a system state which provides the output signal with the periodic dynamic of the input signal. Let us observe the situation shown in Figure 8. The transitions due to interaction between the noise and the waveform (a) are less regular than those linked to the forcing signal (b).

As an extreme case it is evident that a forcing signal with sufficient amplitude to trigger off system transitions autonomously ($A>S$) guarantees that the *periodicity condition* is met.

To account numerically for the link between the amplitude A and the increase in the maximum signal-to-noise ratio, we have to bear in mind that when the output signal tends to become periodic with a period ν, as a consequence of the increase in A, the energy associated with the fundamental harmonic of the output tends to grow, as does the amplitude of the signal spectrum at that frequency. Hence the increase in the numerator of R and therefore of R itself.

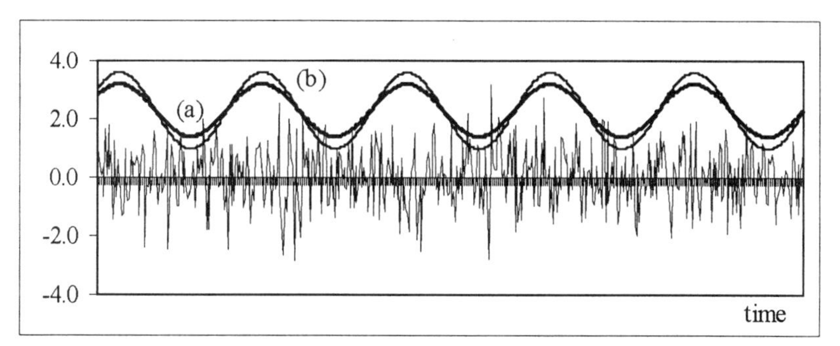

Figure 8. The transitions due to interaction between noise and waveform depend on the forcing signal amplitude.

2.2.5 AN OVERVIEW OF THE *QDW* SYSTEM: STOCHASTIC DRIVING, PERIODIC DRIVING, STOCHASTIC RESONANCE CONDITION

Let us again consider system (2.1), which describes Brownian motion, and characterise it in the *SR* sense.

It can easily be demonstrated that if $a>0$ the system possesses one unstable state and two stable states with the following values for x

$$x_0 = 0; x_\pm = \pm\sqrt{\frac{a}{b}} \tag{2.10}$$

which correspond to the maximum and minimum of the potential $V(x)$.

The difference between the amplitude of the maximum potential $V(x)$ and the level of the two minimum points represents the threshold of the system or the energy barrier for the charge, and is equal to

$$DV = \frac{a^2}{4b} \tag{2.11}$$

From equations (2.10) and (2.11) it emerges that it is possible to modify the form of the potential by acting on parameter a or b. If the a/b ratio is left unaltered, it is only possible to vary the amplitude of the barrier, whereas if a and b are varied independently the distance between the two minimum points also varies.

At this point a brief historical note is in order. System (2.1) has been studied for a number of years and still represents a topic of considerable scientific interest. The model has, in fact, been used as a mathematical model to represent a number of natural phenomena. In its most classical sense, for instance, it represents the movement of a steel beam subjected to electromagnetic stress held pinned to fixed supports. The fixed supports provide the system with non-linear stiffness that can be expressed by the term bx^3. In this case, the physical interpretation of the sign of

the parameter is as follows: if $a>0$ the term in x represents a resistive elastic torque, which is the case of *QDW*, while if $a<0$ it represents an active elastic torque.
For this and any other system of the same or greater complexity, can be impossible to find an analytical solution; however, by using numerical simulations it is possible to observe its behaviour.

- *Noise-driven system behaviour*

Let us now analyse the behaviour of particles trapped in the energy barrier and only subjected to a stochastic forcing signal $\varepsilon(t)$

$$\ddot{x} = -\gamma\dot{x} + ax - bx^3 + \varepsilon(t) \tag{2.12}$$

The statistical properties of the noise $\varepsilon(t)$ can easily be synthesised from its average value and the auto-correlation function, which are

$$\mu_{\varepsilon(t)} \tag{2.13 a}$$
$$C_\varepsilon(t)=2\gamma KT\delta(t) \tag{2.13 b}$$

For a through understanding of the system behaviour when forced by a stochastic signal $\varepsilon(t)$ the following time scales of the system must be taken into account: the noise time correlation τ, the damping relaxation time γ^{-1} and the librational period of the Brownian particle about the potential minima [8, 9]. The spatial spectral density shows no resonance peak in the overdamped limit $\gamma>>\omega_p$ $(\omega_p=2\pi v_p)$. On the other hand, if the underdamped limit (corresponding to the requirement $\gamma<<\omega_p$) is guaranteed, a resonance behaviour is observed [5].
Another characterisation can be based on the potential barrier value. In condition of high potential barrier ($DV>>KT$) the particle is trapped in one of the two wells therefore oscillating inside one of them. On the other hand if $DV<KT$ the particle can oscillate from the left to the right potential well and viceversa.
It is also recalled that the *Kramers Rate* for the *QDW* system can be calculated as follows

$$R_K = \frac{\omega_0\omega_b}{2\pi\gamma}e^{-\frac{DV}{D}} \tag{2.14}$$

where

$$\omega_b=[V''(x_\pm)]^{1/2}=(2a)^{1/2}$$
$$\omega_0=[V''(x_0)]^{1/2}=(a)^{1/2} \tag{2.15}$$

represent the system frequencies associated with the potential minima and maximum.

The steady state relaxation dynamics of a bistable system can be characterised by means of both the power spectral density of the system output $S(\omega)$ and the correlation function

$$C(t)=\langle x(t)x(0)\rangle / \langle x^2\rangle \tag{2.16}$$

These two quantities are linked by the Fourier transform operator as the *Wigner-Khintchine* theorem states.
Figure 9 gives an example of the auto-correlation function, while Figure 10 shows the spectrum at varying values of the parameters γ and DV/KT.
Examining the results of this analysis, we note the presence of two peaks, one due to the vibration (or relaxation) of the particles in a well (librational peak), equal to ω_b, and another to oscillation between the two minimum states (activation peak), equal to $\omega_b/2$. In addition, a low-frequency component can be observed, due to random movement of the particles out of the two wells (escape rate).
Analysis of Figure 10a confirms that it is possible to reduce the amplitude of the librational and activation peaks by increasing γ.
Figure 10b, on the other hand, shows that as the DV/D ratio increases, the librational peak can be emphasised at the expense of the activation peak. These figures show the spectrum with increasing b values and therefore decreasing values for both the threshold DV and the distance between the two wells.

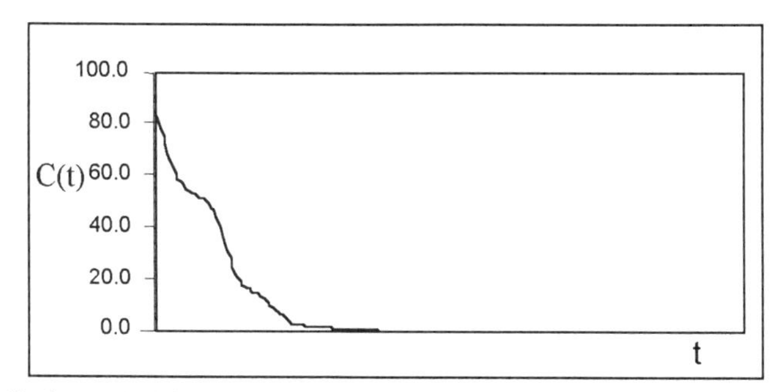

Figure 9. Qualitative trend of $C(t)$ when $a=1$; $b=1$; $\gamma=0.05$.

This behaviour (reduction of the activation peak as b decreases) can be interpreted as incapacity on the part of the particles to oscillate from one state to another as the potential barrier or the inter-well distance increases.

Figure 10c gives the results obtained in a set of simulations run with varying values for the parameter a. In each simulation the parameter b was also varied in such a way as to keep the threshold DV unaltered. As can be observed, the two peaks move in perfect agreement with (2.15). It should be stressed that in practical applications it is not possible to have an ideal delta correlated noise signal that satisfies (2.13). In some cases, for example, the noise signal used is exponentially time-correlated with correlation time τ

$$C_{\varepsilon}(t) = (KT\gamma/\tau)\, e^{-|t|/\tau} \tag{2.17}$$

If the noise $\varepsilon(t)$ is correlated, with the auto-correlation function (2.17), it is necessary to take into account not only the system parameters but also the parameter τ which represents the non-ideal nature of noise. It is demonstrated, however, that when the τ values are very low as compared with the other time constants, the behaviour of the system is again comparable with that due to a delta-correlated stochastic signal.

To highlight the effects of the non-ideal nature of noise, let us consider the spectra shown in Figure 3a when the system is forced by a stochastic component alone. As can be observed, at low frequencies the trend is *lorentzian* and can be analytically formulated with an expression of the following kind

$$S(\omega)\Big|_{\omega << \frac{\omega_b}{2}} = K\frac{\lambda(\tau)}{\lambda^2(\tau)+\omega^2} \tag{2.18}$$

where K is a quantity related to the system parameters, and $\lambda(\tau)$ is the smallest nonvanishing eigenvalue of the *Fokker-Planck* equation associated with the problem being considered [9]. $\lambda(\tau)$ must be related to the rate of escape process out of the potential barrier [8, 9] and it is very close to the *Kramers Rate*. With ideal noise (τ=0), the escape rate is proportional to the *Arrhenius factor*: $\lambda(0) \propto e^{-DV/KT}$.
At a first approximation

$$S(0) \propto 1/\lambda(\tau) \tag{2.19}$$

It is possible to make some interesting considerations about the influence of coloured noise (τ other than zero) on the behaviour of the system.

For large value of τ, the maximum of $S(\omega)$ in $\omega=0$ is enhanced. This implies that the smallest eigenvalue $\lambda(\tau)$ decrease with increasing τ. This shows that the escape rate $\lambda(\tau)$ clearly depends on the characteristics of the noise (τ): the higher the value of τ the lower the escape rate.

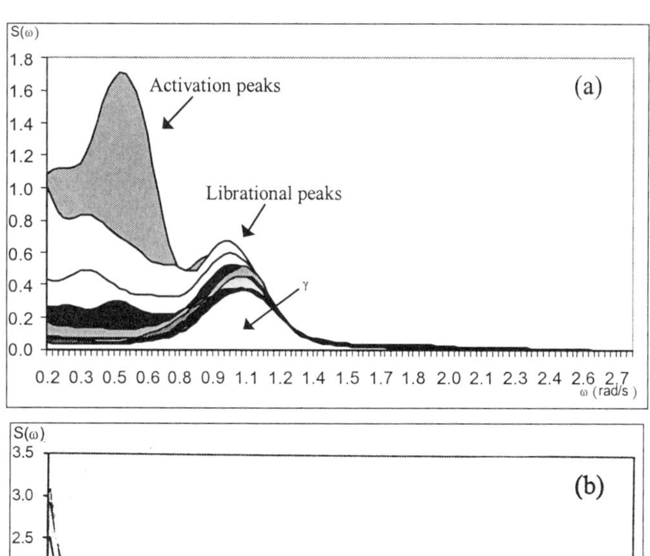

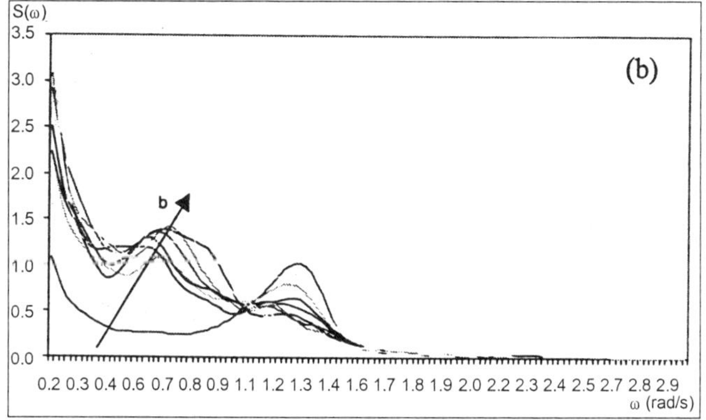

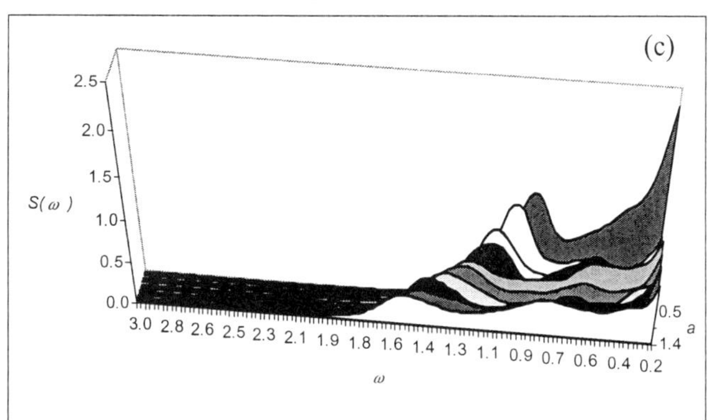

Figure 10. (a) Output signal spectrum when $a=1$; $b=1$; $\gamma=[0.05:0.02:0.15]$. Note the reduction in the two peaks as the damping factor γ increases. (b) $a=1$; $b=[0.5:0.1:1.4]$, $\gamma=0.085$. Note the evolution of the two peaks as the parameter b increases. (c) $a=[0.5:0.1:1.4]$; $b=1$, $\gamma=0.085$. Note the position of the two peaks as the parameter a increases.

In particular with time-correlated noise it has been experimentally proved that $\lambda(\tau)$ decreases exponentially with increasing τ^2, according to the heuristic law

$$\lambda(\tau) = \lambda(0)e^{-k_0\tau^2} \propto e^{-\frac{DV}{KT}} e^{-k_0\tau^2} \tag{2.20}$$

Where k_0 depends on the DV/KT ratio. Hence, unlike the R_K, $\lambda(\tau)$ is no longer proportional to the *Arrhenius factor* $e^{-DV/KT}$. This confirms the influence of coloured noise on the escape rate.

- *The effect of a periodic forcing signal*

If a periodic forcing signal with an amplitude A and a frequency ν is applied to the particle, the potential (2.2) is modulated upwards and downwards. During these fluctuations the threshold DV will undergo positive and negative variations. If the amplitude of the forcing signal is sufficiently higher than the threshold an oscillatory mechanism will be set off that will give the particles a periodicity equal to that of the forcing signal; otherwise they will be confined within one of the two wells.

The system being investigated strongly depends on parametric variations. As the parameters vary, the system modifies its behaviour, passing from periodic states, characterised by limit cycles, to chaotic behaviour [30].

In the case being dealt with in this section, the parameters will be chosen in such a way that the system presents a variety of non-linear dynamics and, according to the value of the forcing signal, its state remains confined in the proximity of a minimum or oscillates between the two minima.

Let us now consider system (2.1), which can be rewritten for the sake of convenience as follows

$$\ddot{x} = -\gamma\dot{x} + ax - bx^3 + f \tag{2.21}$$

By simple calculation we can achieve a system of two first-order equations

$$\begin{cases} \dot{x}_1 = x_2 \\ \dot{x}_2 = -\gamma x_2 + ax_1 - bx_1^3 + f \end{cases} \tag{2.22}$$

The system typically has dynamics at a very low frequency [8, 12, 14, 30]. As one of our aims is to develop an analog simulation environment, we chose temporal scaling of the linear part of the system to vary the system bandwidth.

With this procedure we obtain the following system

$$\ddot{x}/k^2 = \frac{-\gamma}{k}\dot{x} - bx^3 + ax + f \tag{2.23}$$

Hence

$$\begin{cases} \dot{x}_1 / k^2 = x_2 \\ \dot{x}_2 = -\gamma k x_2 - bx^3 + ax_1 + f \end{cases} \tag{2.24}$$

The transfer function between the output $y=x_1$ and the input f, considering the linear part of 2.24, takes the following form

$$G(s) = \frac{k^2}{s^2 + \gamma ks - ak^2} \tag{2.25}$$

To highlight the behaviour of the system with varying forcing signal amplitudes, *Simulink®* simulations can be run. An example of the analytical scheme of the system is given in Figure 11. The set of parameters (a, b, γ) used in the simulations is one commonly found in the literuture.

$\gamma = 0.236$
$a = b = 0.967$

Figures 12-14 show the behaviour of the system when subjected to forcing signals with different amplitudes. It is recalled that the system threshold is

$$DV = \frac{a^2}{4b} \approx \frac{1}{4} \tag{2.26}$$

With amplitudes below the system threshold the system makes no state transition and the state x_1 remains confined around x_-. This is evident from the trend of the state variable x_1 and the state diagrams in Figure 12. The reason why the system state falls into this attractor is to be sought in the value of the initial conditions of the system. For these simulations the initial conditions were set in the attraction domain of x_-.
As the amplitude of the forcing signal grows, a gradual passage from a strange to a periodic regime is observed, the latter having a frequency equal to that of the forcing signal, as illustrated in Figures 13 and 14.

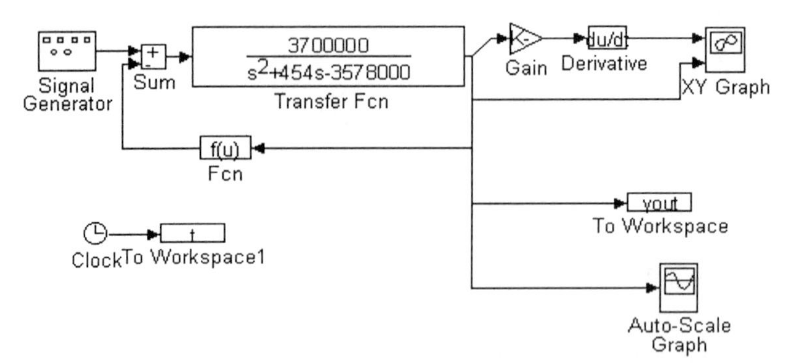

Figure 11. Block diagram implemented in *Simulink®* for *QDW* simulations in the presence of a periodic forcing signal.

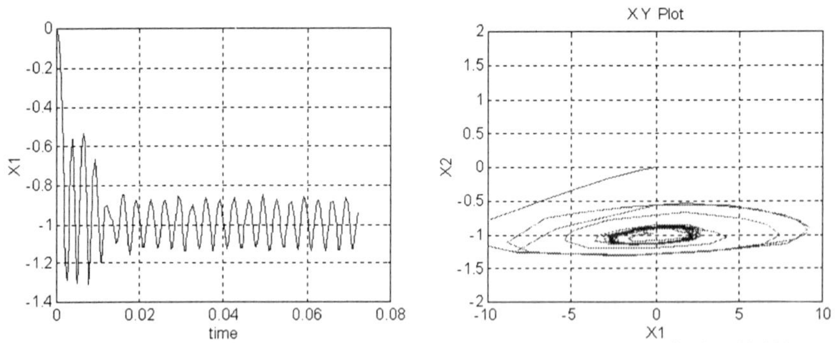

Figure 12. Temporal trend of x_1 and system state plot, with a forcing signal amplitude of 0.1V.

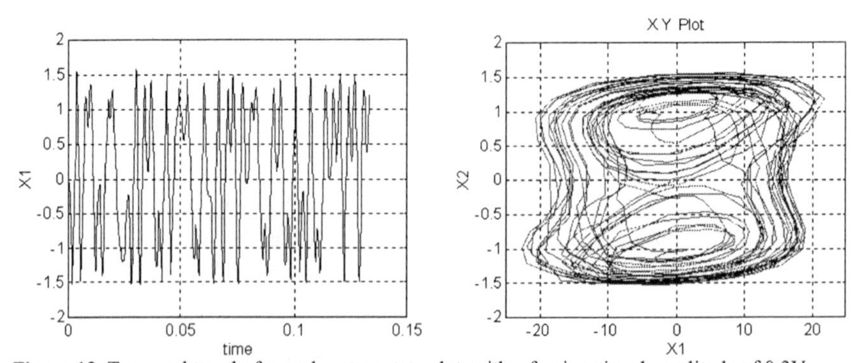

Figure 13. Temporal trend of x_1 and system state plot, with a forcing signal amplitude of 0.3V.

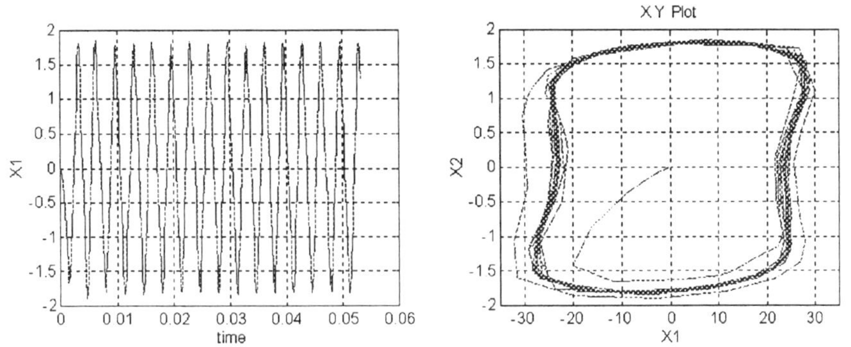

Figure 14. Temporal trend of x_1 and state space, with a forcing signal amplitude of 0.5V.

Besides observing the behaviour of the system with varying forcing signal amplitudes, it is interesting to investigate, as was done with exclusively stochastic signals, the system's response when the parameters a and b are varied. Simulations were therefore run varying the parameters A, ν, a and b. The calculation environment was again *Simulink®*, called up cyclically by a *Matlab®* application. With this strategy the simulation process was made automatic and at the same time it was possible to carry out a cross investigation with *uniform* increases in the parameters being considered.

The variation intervals used in each simulation were as follows

A=[0.1:0.05:0.5] V
ν=[0.4:0.4:2] rad/s
a=[0.1:0.6:6]
b=[0.1:0.6:6].

In the first set of simulations A, ν and a were varied and b was chosen in such a way that DV remained constant and equal to 0.1.

The aim of these simulations was to see how it was possible, once A and ν (and therefore the forcing signal) are fixed, to reach an optimal switching condition (emphasising the action of the forcing signal) and how this is linked to the choice of an optimal value for the parameter a.

By way of example, Figures 15 and 16 show the *QDW* output signal spectra for two pairs of values (A, ν) with varying values for the parameter a.

As can be seen, the family of spectra presents a peak corresponding to the frequency of the periodic forcing signal.

In order to identify the value of the parameter a with which the optimal switching condition occurs, a selection algorithm was used based on the following criteria:

1) if with a certain parameter a value the system output switches between the two states x_- and x_+, the a value is given an index $q=1$; otherwise $q=0$;

2) if the output signal spectrum has a maximum corresponding to the frequency of the forcing signal and if $q=1$, then an index n is increased by one unit. The value of the index n associated with each pair (A, ν) indicates the number of a values that make the system capable of switching between one state and another at the frequency of the forcing signal;

3) in all cases in which $n>1$, the index $J=M_2/(M_1+M_2+M_3)$. Is calculated, where M_2 represents the maximum value of the spectrum at the frequency of the forcing signal, and M_1 and M_3 are the upper value bounds closest to it. The index J therefore weights the amplitude of the maximum point of the spectrum in relation to the other peaks, thus showing the system's capacity to outline the presence of the periodic input component;

4) the a value at which the index J is maximum is calculated and taken as the optimal parameter.

The optimal parameter a values for each amplitude A and pulse ν are given in the map in Figure 17. An interesting property is observed: as the frequency increases, when the threshold is kept constant, the optimal value for a increases. It is possible to demonstrate that this corresponds to reduce the distance between the two wells at increasing frequency values. For the threshold to remain constant and equal to K, in fact, the parameter b has to take the value $b=a^2/(4K)$; the distance between the two wells therefore becomes $d=2(a/b)^{1/2}=4(K/a)^{1/2}$, so as a increases the distance d decreases.
It is also observed that when the frequency is kept constant, as the forcing signal amplitude A increases the optimal value for a decreases. It is possible to demonstrate that this corresponds to increase the distance between the two wells when the amplitude of the periodic forcing signal increases.
Figure 18 shows the trend for the index J with varying A and ν. As can be seen, at low amplitudes as the frequency increases the index $J(A, \nu)$ gradually decreases, reaching zero when $A=0.1$V and $\nu=2$ rad/s. At the same time, in the remaining area its value is close to one.
Another interesting result is given by the trend of the index $n(A, \nu)$, shown in Figure 19. Parameter n increases as the forcing signal amplitude A increases. This can be accounted for by an increase in the transition probability which, as pointed out previously, increases along with A.
A second set of simulations was run to observe the system's behaviour when the parameter b is varied. The aim of these simulations was to see, once A and ν (and

therefore the forcing signal) are fixed, how it is possible to reach an optimal switching condition (emphasising the action of the forcing signal) and how this is correlated with the choice of an optimal value for b.

Figure 20 shows the *QDW* output signal spectra for a pair of values (A, ν) with varying b values. As can be seen, here again the spectra present a peak corresponding to the frequency value. In order to identify the b value at which the optimal switching condition is achieved, the same selection algorithm can be used.

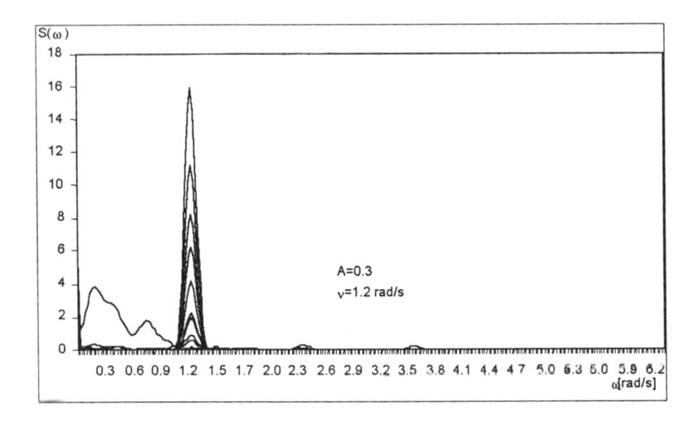

Figure 15. Output signal spectra when $A=0.3$ *V* e $\nu=1.2$ *rad/s*, for different values of the parameter a. The curves show a peak correspondent to the forcing signal frequency.

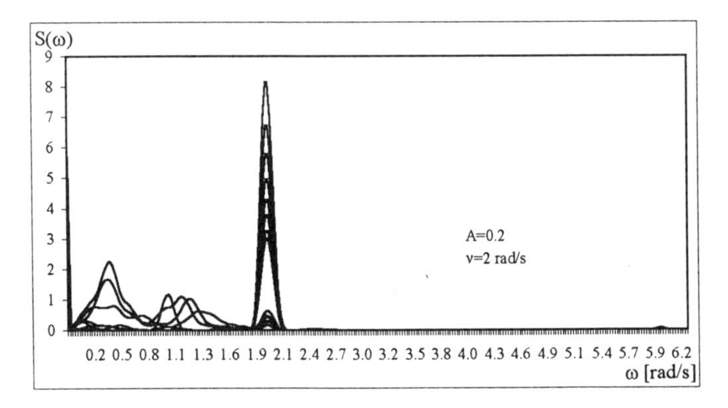

Figure 16. Output signal spectra when $A=0.2$ *V* e $\nu=2$ *rad/s*, for different values of the parameter a. The curves show a peak correspondent to the forcing signal frequency.

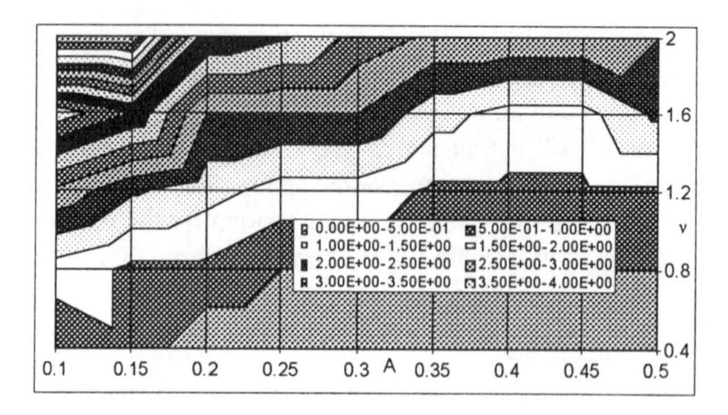

Figure 17. Map of the optimal values of the parameters a.

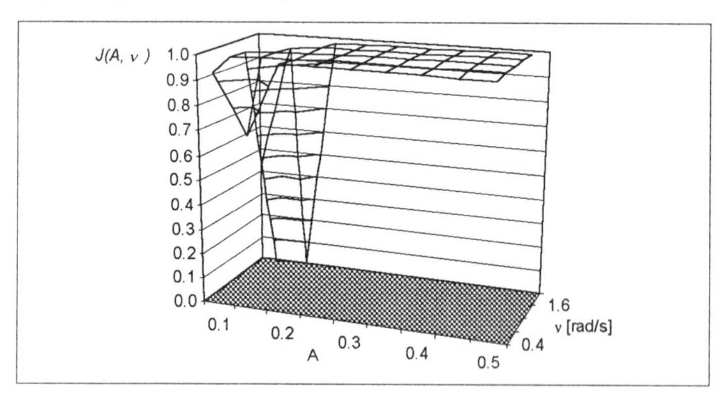

Figure 18. The selection index $J(A, \nu)$.

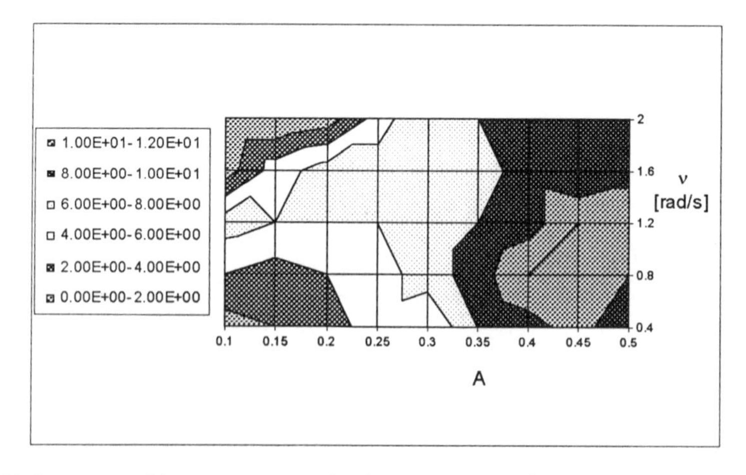

Figure 19. Occurrences of the parameter a assuring the system commutations.

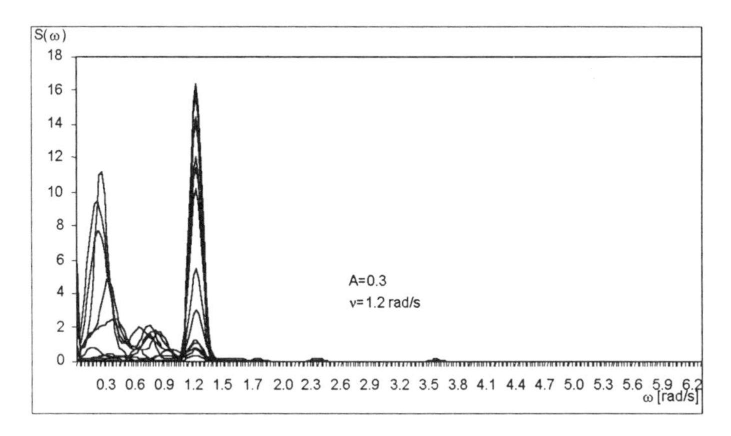

Figure 20. Output signal spectra when $A=0.3$ *V* and $\nu=1.2$ *rad/s*, for different values of the parameter b. The curves show a peak correspondent to the forcing signal frequency.

- *System forced by a periodic and a stochastic component.*

The results presented in this section concern a system forced by a periodic component to which noise with known characteristics is added. The stochastic signal used has a *Gaussian* probability density with a null mean value and a standard deviation of σ. Cases dealt with up to now in the literature confirm that the action of a stochastic component enables the system to switch even when the amplitude of the forcing signal is lower than the threshold. This happens with optimal noise amplitude values. Figures 21-23 show the evolution in time of the system state and its spectral components in three cases:

- transition caused by the periodic forcing signal;
- failed transition due to insufficient forcing signal amplitude;
- transition caused by the same forcing signal in case 2 with the addition of noise with a suitable amplitude.

For a given period of the forcing signal T, the matching condition can be fulfilled by tuning the noise level D to the value determined by Equation (2.7) This mechanism is discussed in detail in [20].
Tuning the noise variance is just one way to fulfil Equation (2.7). It should be observed that, when the characteristic of both the forcing signals $f(t)$ and $e(t)$ are fixed, the system parameters a and b can be tuned to obtain synchronism between the transition rate and the period of the forcing signal.
In Chapter 3 a new methodology based on this feature is proposed, in order to obtain spectral information from a periodic signal modulated by large amplitude noise.

The main idea investigated here is the possibility to move a system into the *SR* condition when it is forced by an unknown periodic signal modulated by a white *Gaussian* noise with unknown statistical properties, by changing its structural parameters a and b. It should be noted that in this case the damping factor γ should be kept low in order to run the underdamped mechanism in the system.
If the system (2.1) is forced by a periodic signal and noise the waiting time between two transitions can be monitored. The distribution of this quantity shows several peaks decreasing in amplitude. The first peak corresponds to the forcing frequency ν while the other peaks are odd multiples of ν. The area E under the first peak corresponds to the probability that the system is synchronised with the periodic signal. Of course the value of E changes according to the system parameter considered. For example if noise variance is considered, E will reach a maximum value when $\sigma = \sigma_{opt.}$ Similar considerations can be made on the possibility of synchronising the system by acting on the parameters a and b, the other quantities being fixed. Hence, for a suitable range of these parameters the waiting time distribution will show a maximum corresponding to the forcing frequency and the area E under this maximum will be magnified with a optimal value of a and b.
Another way to move the non linear system into the *SR* condition could be to tune the forcing signal frequency. It is interesting to observe that if the quantity E is plotted as a function of the forcing frequency ν it shows typical resonant behaviour, thus confirming that when all the stochastic system parameters (noise variance, forcing amplitude, structural parameters) are fixed a resonant frequency can be defined [26].
On the basis of the above consideration the following rules can be obtained:

- for suitable range of the stochastic system parameters, the power spectrum of the system output will show a peak corresponding to the forcing frequency;
- this peak will reach maximum value if the values of the parameters are optimal.

The last task explored in this section is to show the possibility of satisfying the matching condition by adapting the system to the input signal and not vice versa. Therefore, the characteristics of the forcing signal being fixed, optimal values for the parameters a and b will be searched for.
The procedure for detection of system parameters allowing system synchronisation starts with computation of the system output for a large set of values of a and b. The algorithm computes the frequency corresponding to the peak value for each spectrum, computes the distribution of these frequencies and chooses the frequency with the maximum number of occurrences. It has been experimentally proved that this maximum value is reached by the forcing frequency.

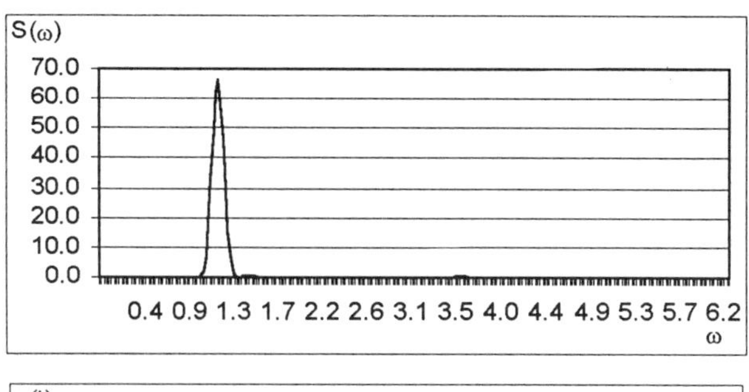

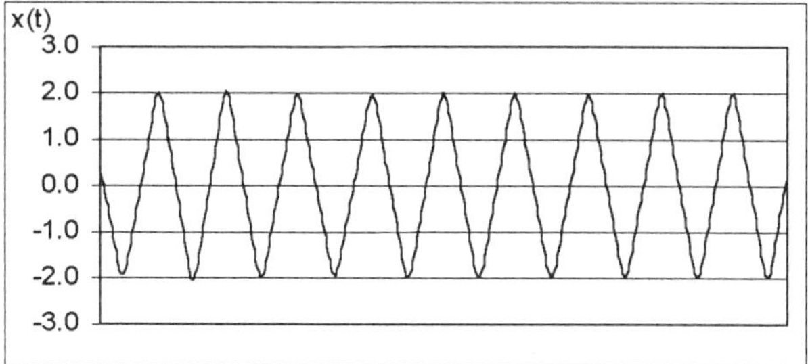

Figure 21. Evolution in time of state of system and its spectral components: transition due to periodic forcing signal (*a=b=1; γ=0.085; ν=1.2 rad/s; A=0.5*).

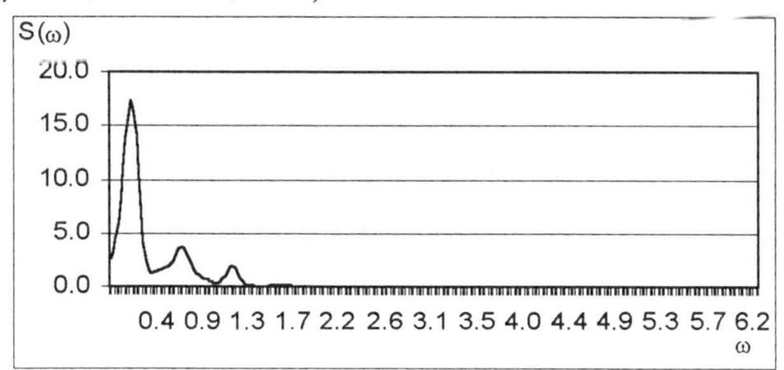

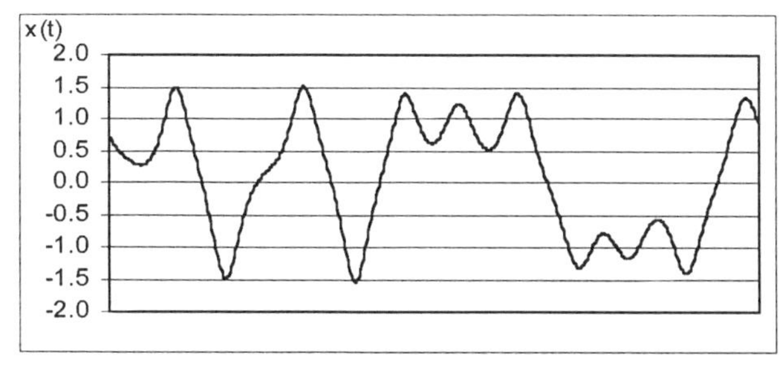

Figure 22. Evolution in time of state of system and its spectral components: failed transition due to insufficient forcing signal amplitude (*a=b=1; γ=0.085; ν=1.2 rad/s; A=0.2*).

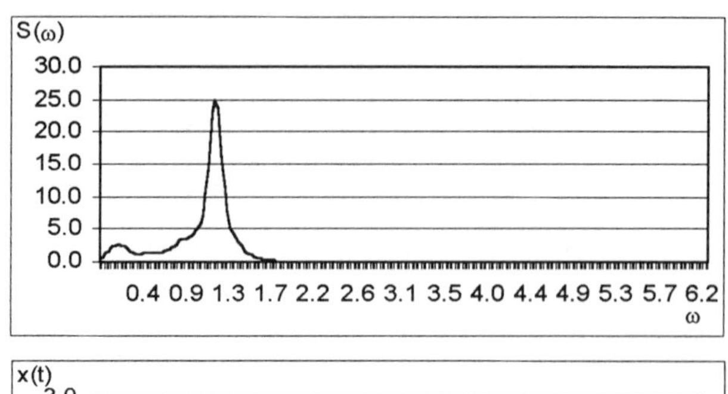

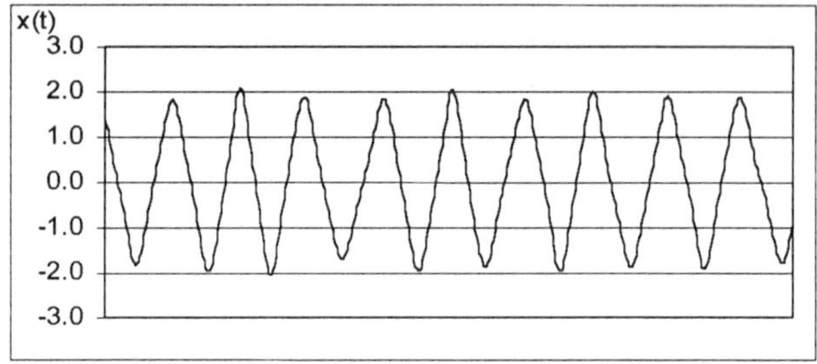

Figure 23. Evolution in time of state of system and its spectral components: transition caused by same forcing signal with the addition of noise with a suitable amplitude ($a=b=1$; $\gamma=0.085$; $\nu=1.2$ *rad/s*; $A=0.2$).

Let us assume, for example, that the system is forced by a signal made up of a harmonic component modulated by a stochastic series of sufficient amplitude to mask the former.

Figures 24-29 show:

a) the spectra of the forcing signal;
b) the spectra of the system response when parameters a and b are varied;
c) the spectra of a reduced set of responses after application of the selection algorithm.

Let us analyse Figure 24. From the spectrum of the forcing signal, in the top section of the figure, it is not possible to extract information about the frequency of the modulated periodic component. The middle section shows the spectra of the *QDW* system output for a large set of structural parameters. In the bottom section the spectra of the output showing the maxima in a narrow range of the frequency selected by the algorithm are presented.

To check the usefulness of the selection algorithm a classical analysis of the input signal was carried out. Examples of the correlation map and the statistical distribution of the signals are given in Figures 30-35. It can be observed that by using the mentioned approaches no information can be obtained.

A useful environment for noisy signal filtering will be discussed in detailed in Part 2.

For the time being, our attention is focused on the usefulness and advantages of knowledge of noise-added phenomena and their application to concrete cases and real problems. The *QDW* example and the research that has been done into it only represent one in a wide range of problems to which the use of noise addition techniques provides an interesting solution.

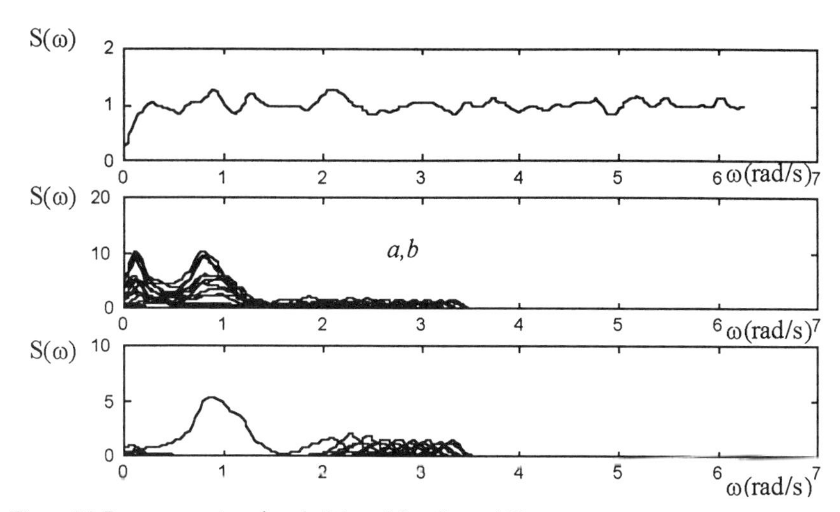

Figure 24. Response spectra when $A=0.1$, $\nu=0.8\ rad/s$, $\sigma=1\ V$.

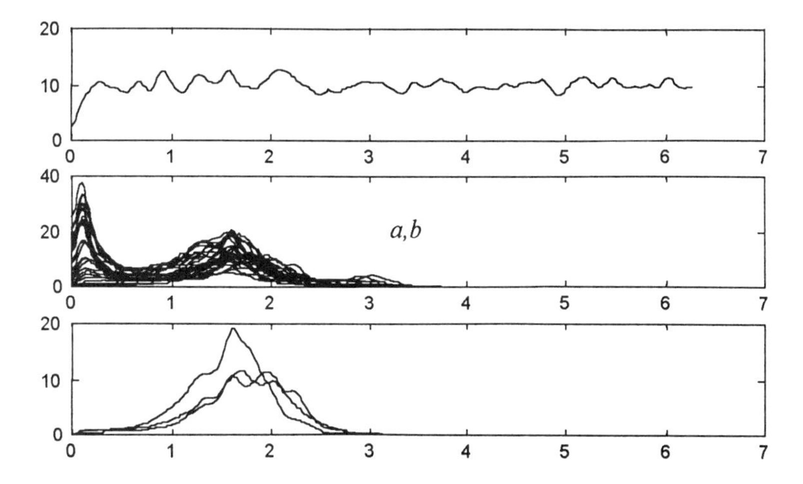

Figure 25. Response spectra when $A=0.4$, $\nu=1.6\ rad/s$, $\sigma=3V$.

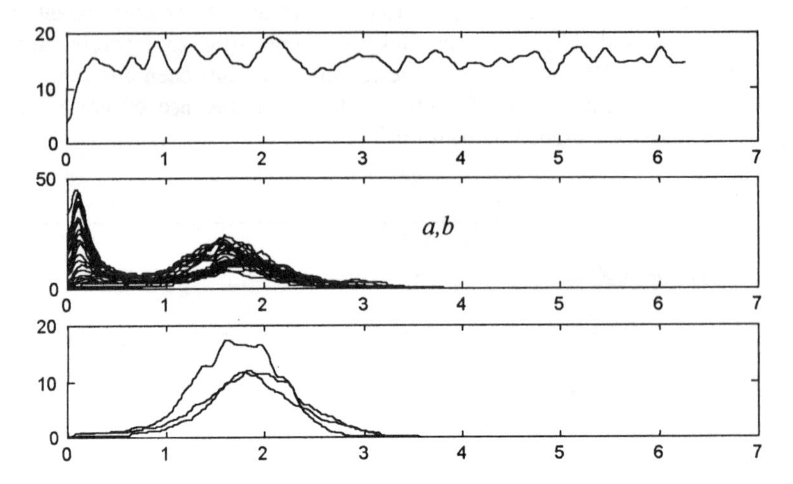

Figure 26. Response spectra when $A=0.4$, $\nu=1.6$ *rad/s*, $\sigma=4V$.

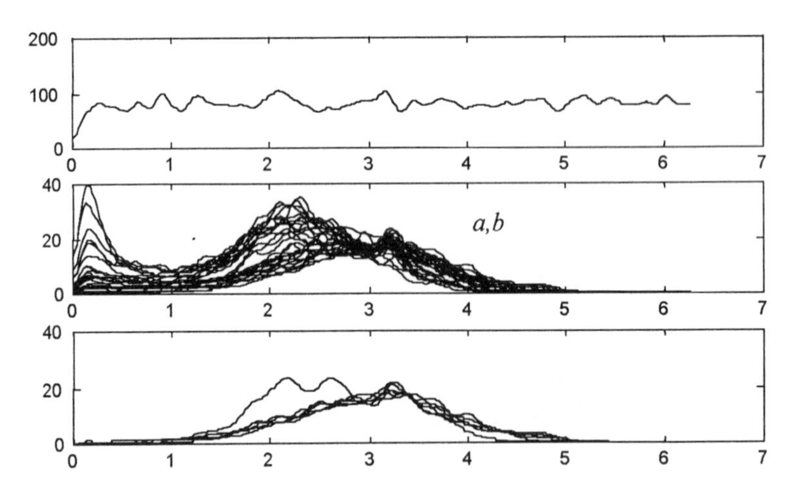

Figure 27. Response spectra when $A=1.2$, $\nu=3.2$ *rad/s*, $\sigma=28V$.

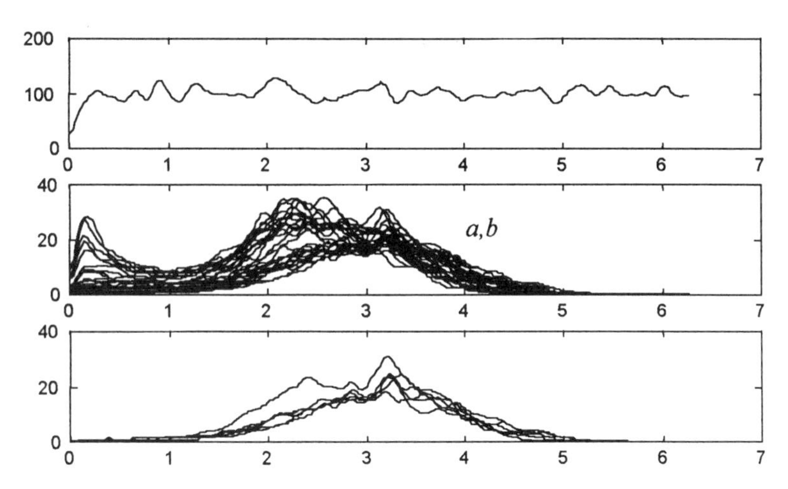

Figure 28. Response spectra when $A=1.2$, $\nu=3.2$ *rad/s*, $\sigma=10V$.

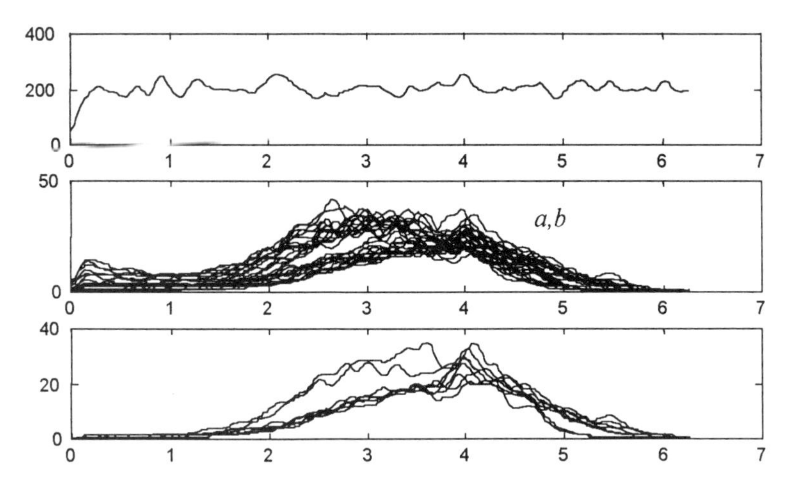

Figure 29. Response spectra when $A=1.8$, $\nu=4$ *rad/s*, $\sigma=14V$.

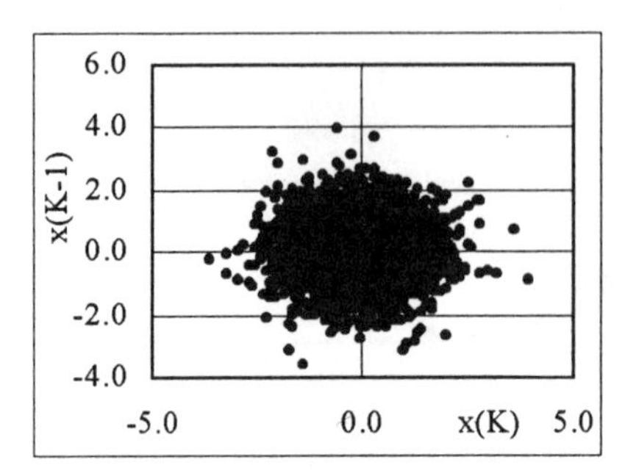

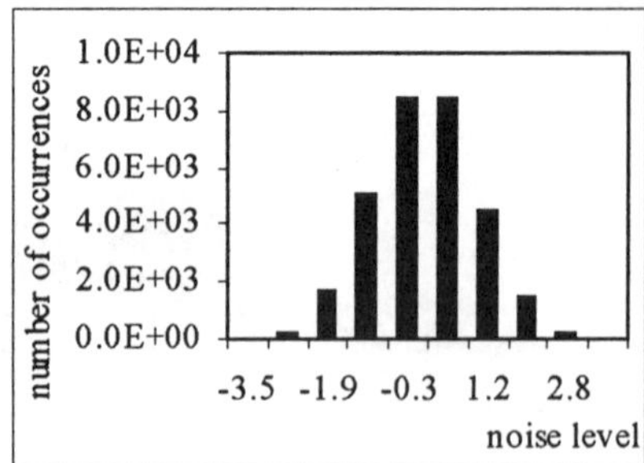

Figure 30. Analysis of forcing signal when $A=0.1$, $\nu=0.8$ rad/s, $\sigma=1V$.

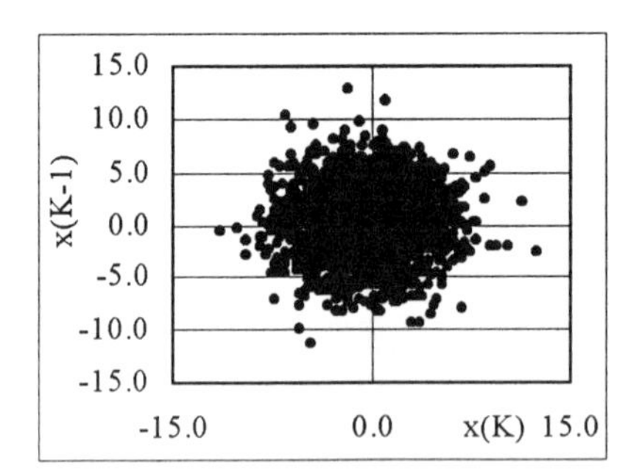

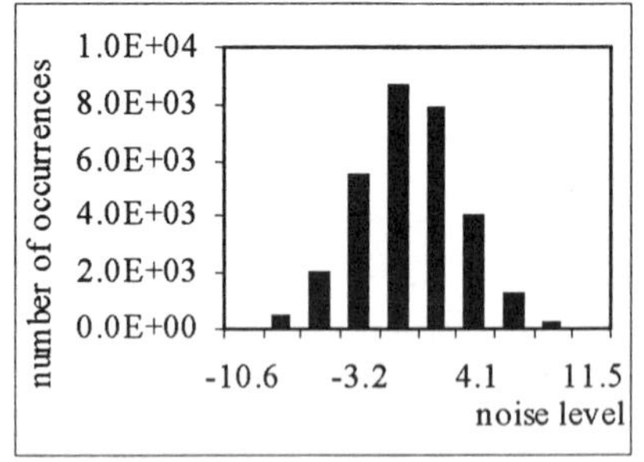

Figure 31. Analysis of forcing signal when $A=0.4$, $\nu=1.6$ rad/s, $\sigma=3V$.

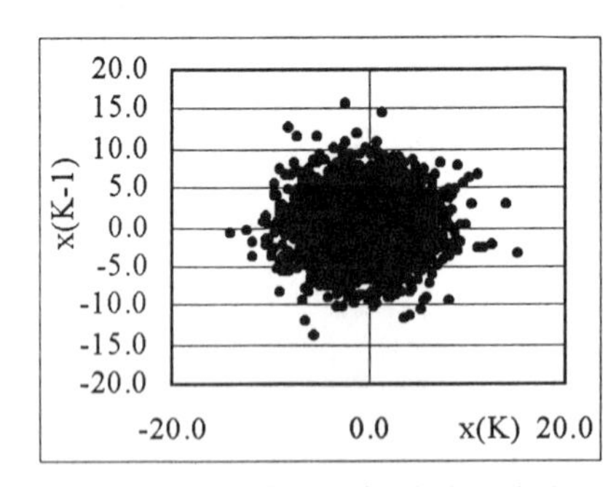

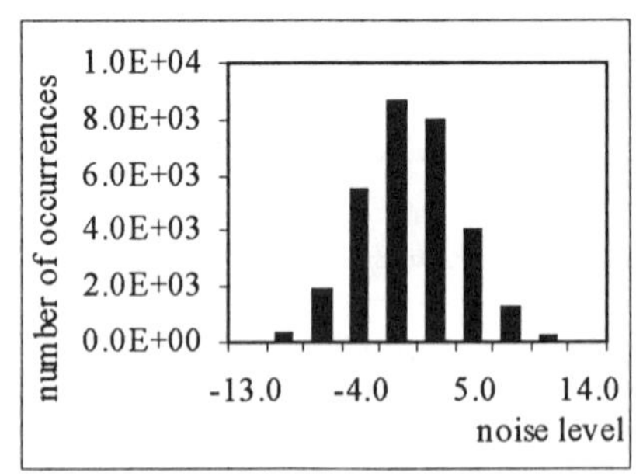

Figure 32. Analysis of forcing signal when $A=0.4$, $\nu=1.6$ rad/s, $\sigma=4V$.

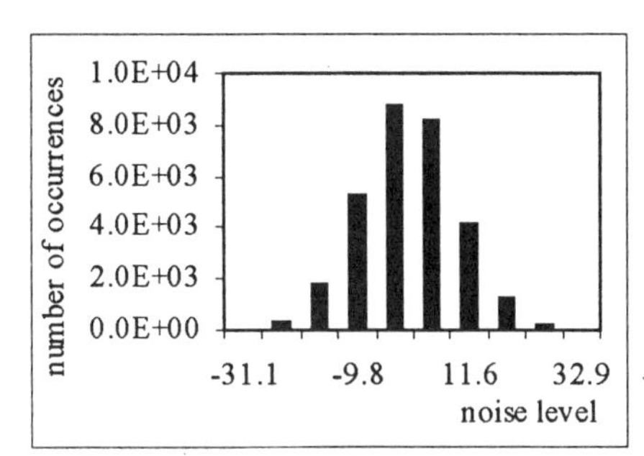

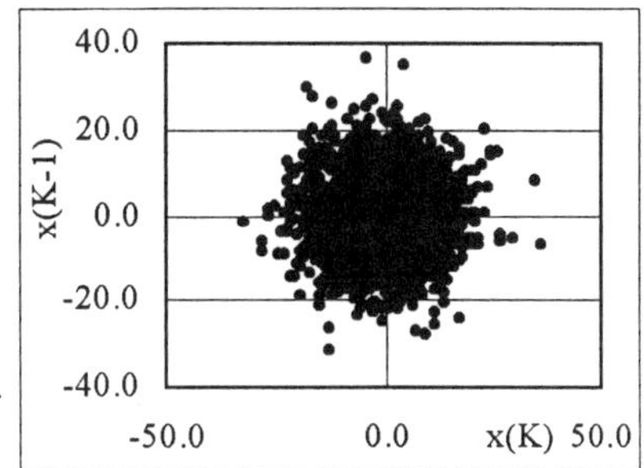

Figure 33. Analysis of forcing signal when $A=1.2$, $\nu=3.2$ rad/s, $\sigma=28V$.

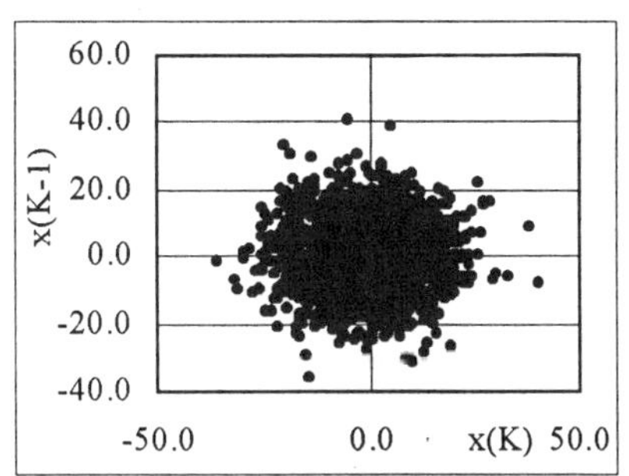

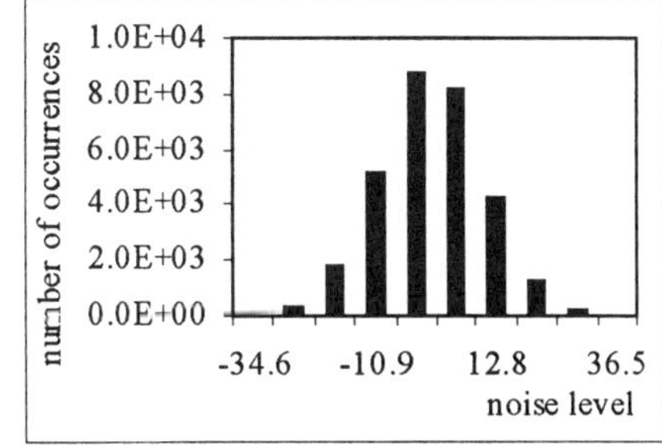

Figure 34. Analysis of forcing signal when $A=1.2$, $\nu=3.2$ rad/s, $\sigma=10V$.

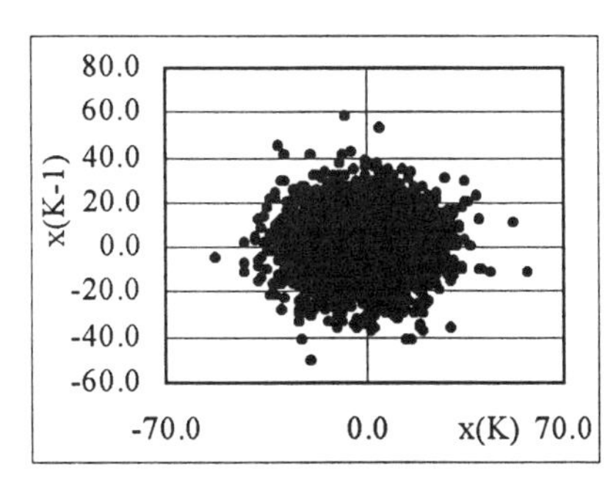

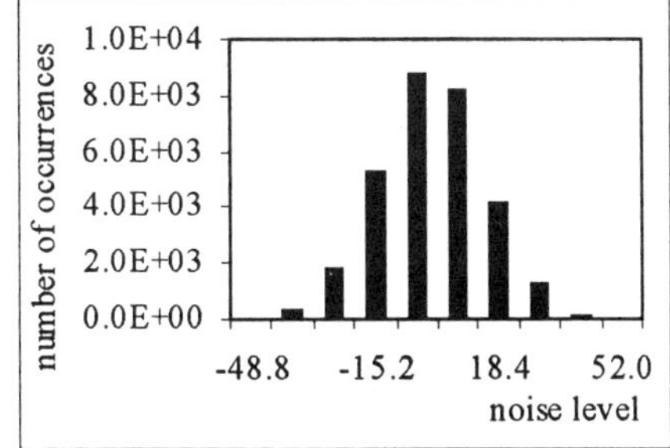

Figure 35. Analysis of forcing signal when $A=1.8$, $\nu=4$ rad/s, $\sigma=14V$.

2.3 Stochastic Resonance in multistable systems

It has been demonstrated that considerations similar to those made regarding bistable systems also hold for multistable systems. Let us consider, for example, an A/D converter, which is a prototype of a multistable system. Observing the trend followed by the spectrum, P_y, of the output signal at the same frequency as the input signal, the following forms of behaviour are recorded depending on the noise variance:

- with a *uniform* noise source, P_y presents peaks corresponding to the different system thresholds;
- in the presence of additional noise with a *Gaussian* distribution, the value of P_y is practically constant as σ varies [19].

The ratio between the value of σ at which P_y reaches its maximum value and that of the corresponding threshold, remains constant. This suggest the possibility of considering multistable systems as a generalisation of bistable systems. It is recalled, in fact, that in bistable systems the signal-to-noise ratio, as a function of σ/S, does not vary when S varies.
It should be pointed out that the phenomena discussed have nothing in common with the dithering techniques used to increase the resolution of an A/D converter. We will discuss these techniques later on.

2.4 Simulation tools for Stochastic Resonance observations

In order to observe the behaviour of an electronic system operating in added-noise conditions, it is possible to use a *PSPICE®* simulator or a numerical simulator such as *Matlab®* or *Simulink®*. In this section we will illustrate the behaviour of a bistable system typically used in studies on *SR* phenomena: the Schmitt trigger. The I/O characteristic of the system is shown in Figure 36.
As can be seen, when the input signal is increased up to $V1$, the output signal goes from $Vmax$ to $-Vmax$, whereas when the input signal is decreased to $V2=-V1$ the output signal passes to $Vmax$.

- Analog simulation

The environment used to carry out initial tests of the system being investigated was *SPICE®*.
When white noise with a null mean and unit variance is applied to the circuit, the systems switches when the value of the stochastic signal is such as to exceed the threshold. As seen previously, spectral analysis shows a power spectrum with a *lorentzian* trend. The circuit used is shown in Figure 37. The system response to the stochastic signal and the system's spectrum are shown in Figures 38 and 39 respectively.

Let us now assume that we apply a sinusoidal signal with an amplitude A and a frequency ν to modulate the system threshold and that the variance σ_2 of the stochastic signal is decreased.
The characteristic parameters of the two signals were chosen in such a way that neither of them was autonomously capable of causing the system to make a transition.
Due to the periodic forcing signal, the threshold is modulated according to a periodic law. At the end points reached in the two semi-periods the situations illustrated in Figure 40 will occur. In these conditions, if the periodic signal is of suitable amplitude, interaction with the stochastic component will enable the system to switch from one state to the other.
This phenomenon occurs wit the periodic transition sequence illustrated in Figure 41. As can be seen from Figure 42, the spectrum of this signal presents a peak at the frequency of the periodic forcing signal. The behaviour observed is typical of a system operating in *SR* conditions. A Schmitt trigger is a system that is often used to study *SR* phenomena and will be referred to frequently in subsequent parts of the volume.

- *MATLAB®* simulation

As an alternative to *SPICE®* simulation, it is possible to study the behaviour of noise-added systems using a numerical calculation environment. In Chapter 6 a tool developed in the *Matlab®* environment with which it is possible to observe the evolution of a system when all its parameters are varied, will be presented.

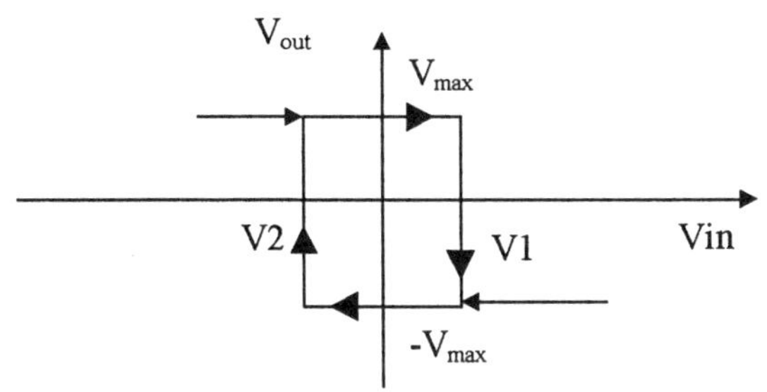

Figure 36. A Schmitt trigger system characteristic.

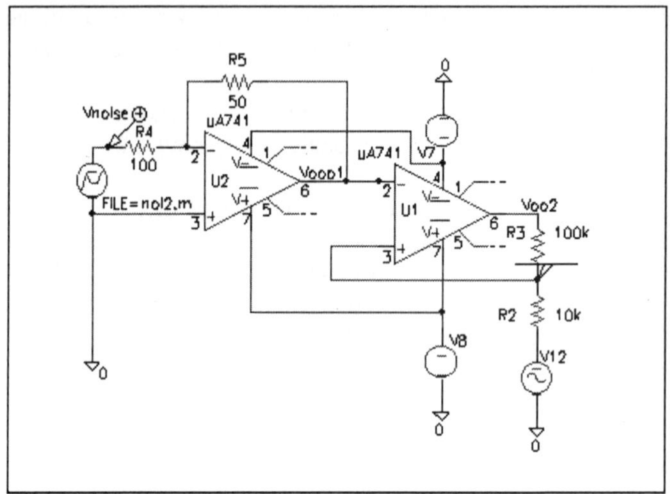

Figure 37. The circuit employed for the Schmitt Trigger implementation.

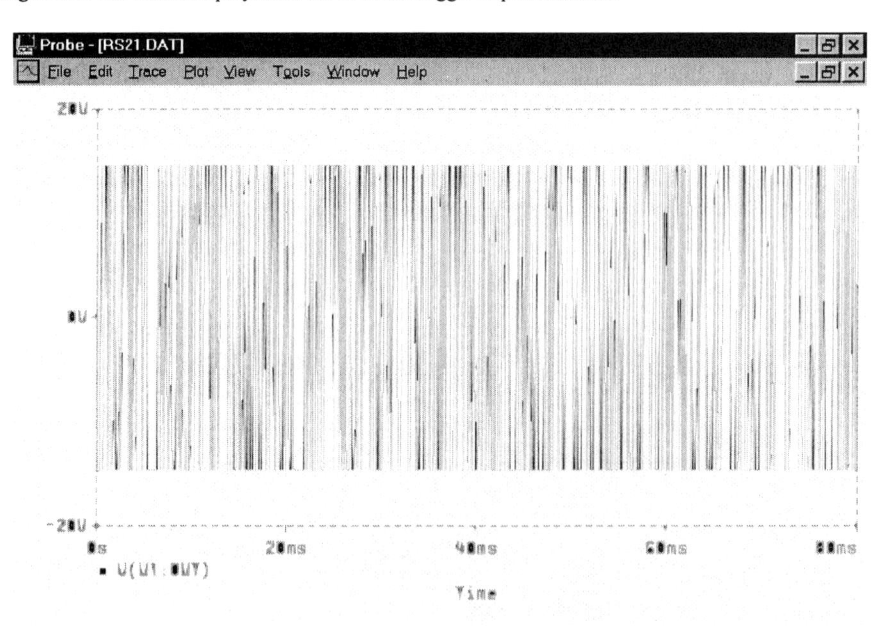

Figure 38. The system output when a noise forcing signal is used.

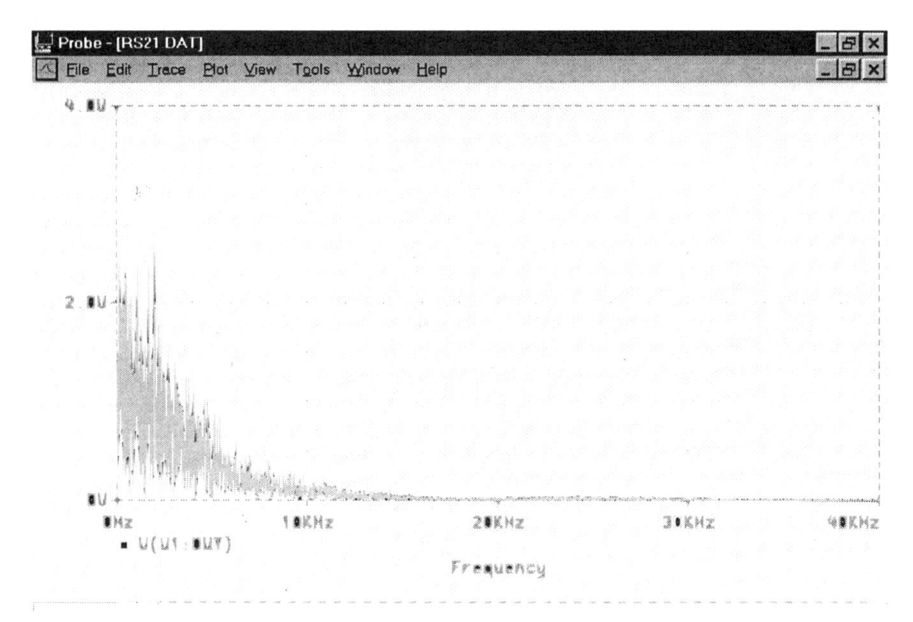

Figure 39. The spectrum of the signal shown in Figure 38.

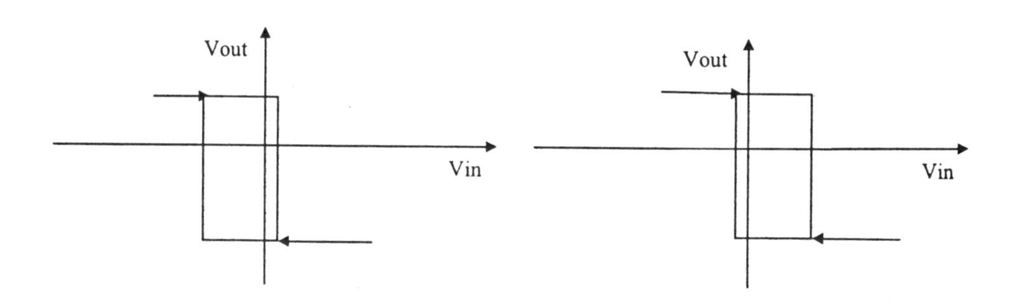

Figure 40. The system behaviour when a periodic signal is modulating the threshold.

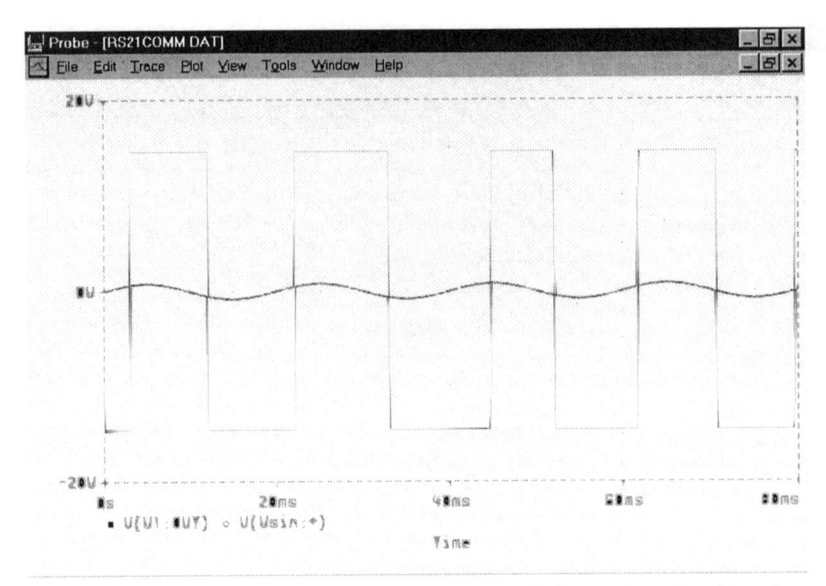

Figure 41. The trigger output when a noise signal and a periodic signal are forced into the system.

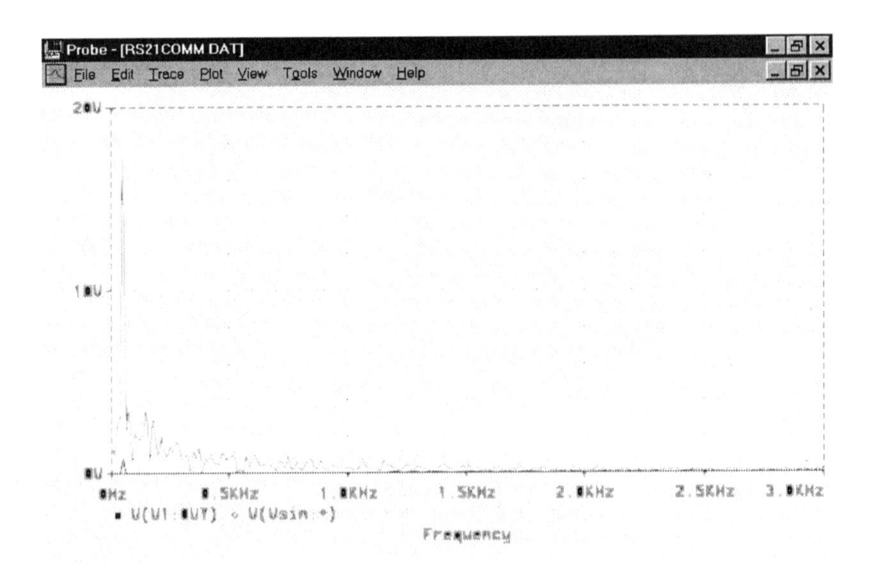

Figure 42. The spectrum of the signal shown in Figure 41.

2.5 Digitisation and dithering

Computers represent analogue signals by using a series of binary numbers which are stored, processed and successively presented to the user. The mapping of a continuous-time, continuous-amplitude signal to a sequence of finite bit-length numbers requires discretisation in both time and amplitude. The processes of discretisation are known as *sampling* and *quantisation* respectively, while the whole process is known as Analog-to-Digital (A/D) conversion.
Physical devices that perform both processes are now widely used and are known as samplers (the simplest form is the *sample and hold* device) and analog-to-digital (ADC) converters.
A example showing a suitable continuous signal and the corresponding vector of samples is given in Figure 43.

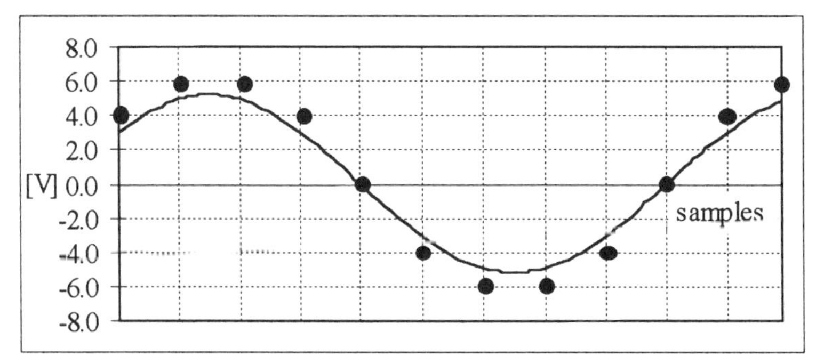

Figure 43. A continuous signal and the corresponding digital representation (after the sampling and quantisation).

The sampling process has been widely studied and understood and a well-defined theory of the sampling process exists. In particular it is known that the sampling operation can be considered as a modulation process of the original signal with a impulse carrier. The spectrum of the sampled signal therefore consists of repetition of the spectrum of the original signal along the frequency axis, the distance between the repetitions depending on the sampling frequency. The *sampling theorem* shows that when the sampling frequency is at least twice the maximum frequency contained in the signal spectrum, the repeated spectra do not overlap and the original signal can be restored. When the hypothesis of the sampling theorem are satisfied, the sampled signal contains the same information as the continuous-time signal.
Unfortunately quantisation is far less understood, mainly due to its non-linear nature. Generally speaking, quantisation is assumed to introduce an error with respect to ideal sampling and number of techniques have been developed to reduce these negative effects. Dithering is by far the most widely-used technique.

A schematic representation of an A/D converter is given in Figure 44. The A/D block can be further schematised by using an ideal sampler followed by a quantiser as shown in Figure 45.

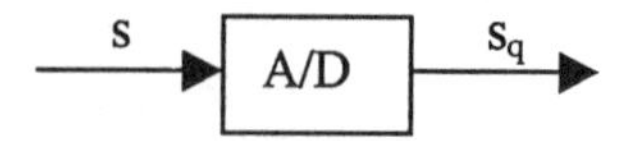

Figure 44. Schematic representation of an A/D converter.

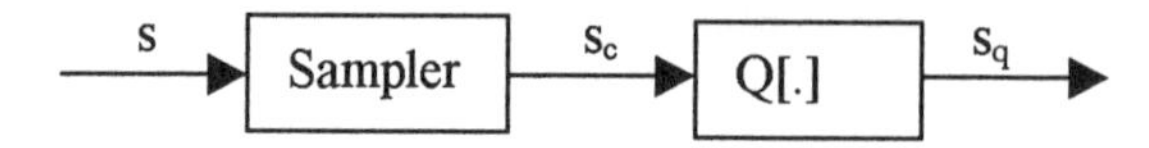

Figure 45. Block diagram of an A/D convereter.

Let us assume that data are represented by using B-bit length representation and that the continuous signal spans the interval [-0,5 A, 0,5 A]. The quantisation step is therefore given by

$$Q = A\,2^{-B} \tag{2.27}$$

Due to the fixed number of bits that are used to represent the original signal (typically eight to sixteen bits), a rounding of the signal occurs. The effects of this rounding process are generally studied by using the so-called *noise model* of quantisation. In particular, the following *quantisation error* is introduced

$$n_q = s_q - s_c \tag{2.28}$$

For the meaning of the quantities in (2.28) see Figure 45 above. Figure 46a shows the characteristic of a *uniform* quantiser, while the corresponding quantisation error is shown in Figure 46b. The quantised signal and the quantisation error can be written as

$$s_q = \left\lfloor \frac{s + \frac{Q}{2}}{Q} \right\rfloor Q \tag{2.29}$$

$$q = s_q - s_c = \left\langle \frac{s + \frac{Q}{2}}{Q} \right\rangle Q \tag{2.30}$$

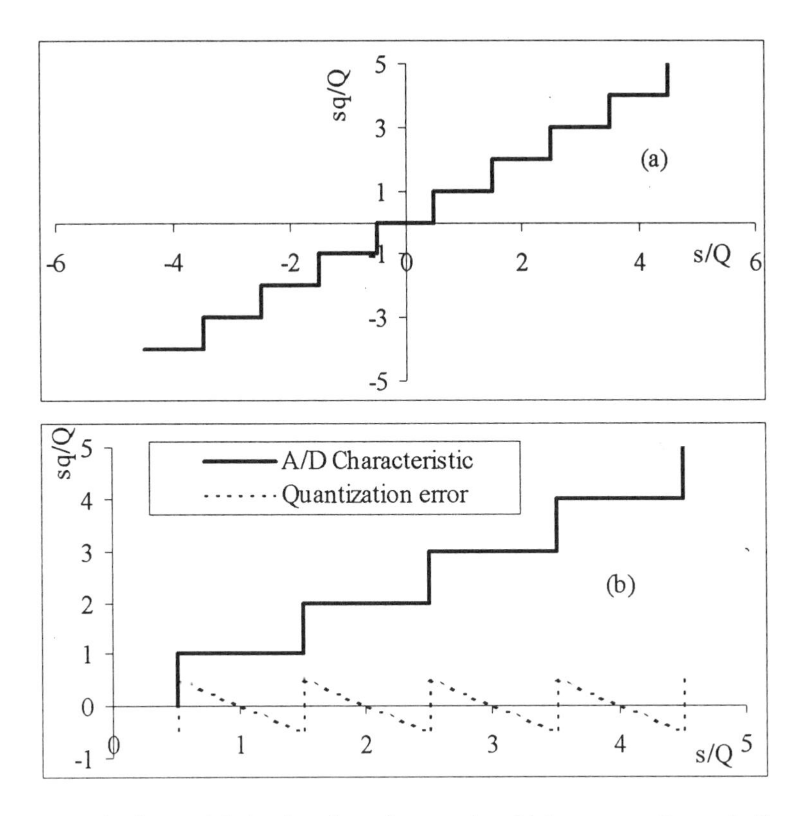

Figure 46. (a) The characteristic function of a *uniform* quantiser; (b) the corresponding quatisation error.

where

$\lfloor \bullet \rfloor$ the integer part operator that gives '*the greatest integer less than or equal to...*';

$\langle \bullet \rangle$ is the fractional operator that gives '*the fractional part of...* '.

Analog-to-digital conversion can be represented as in Figure 47. The figure represents an ideal sampler (this operation does not introduce any error), while the quantisation error is modelled by using an *additive quantisation noise q*, consisting of a sequence of uniformly distributed random variables that are uncorrelated with each other and with the input signal.

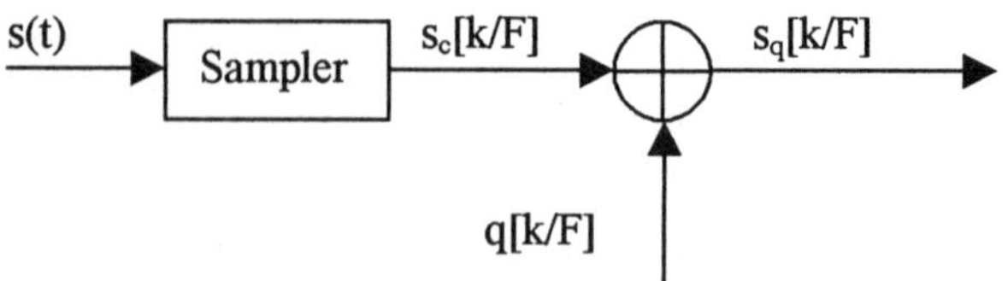

Figure 47. Noise error model of A/D conversion.

This assumption actually depends on the statistical properties of the continuous signal (they can be considered satisfied as long as the number of quantiser levels is large, the quantisation step is small and the input *PDF* is smooth). If more specifically, the input signal can be modelled as a random process whose *PDF* is constant in the interval [-0,5 A, 0,5 A], the quantisation error is a white noise with a zero mean and its *PDF* is uniform in the interval [-0,5 Q, 0,5 Q]. The *PDF*s of both the input signal and the quantisation error are given in Figures 48a and 48b respectively.

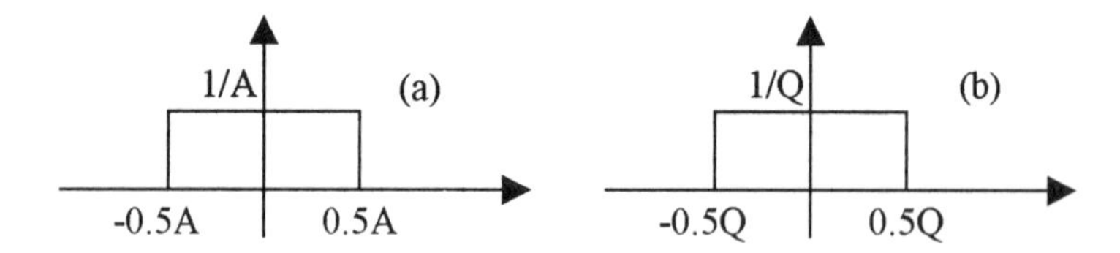

Figure 48. (a) The *PDF*s of the input signal; (b) and the quantisation error.

The noise variance is

$$\sigma_q^2 = \int_{-\infty}^{+\infty} (q - \overline{q}) f(q) dq = \int_{-q/2}^{+q/2} (q) f(q) dq = \frac{Q^2}{12} \tag{2.31}$$

where $\overline{q} = 0$ is the quantisation error mean value and $f(q)$ is its *PDF*.
Clearly, the signal variance is

$$\sigma_{s_c}^2 = \frac{A^2}{12} \tag{2.32}$$

From (2.27),(2.31) and (2.32), the noise error variance is

$$\sigma_q^2 = \frac{Q^2}{12} = \frac{1}{12} A^2 2^{-2B} = \sigma_{s_c}^2 2^{-2B} \tag{2.33}$$

and the number of bits B of the converter is

$$B = \frac{1}{2}\log_2 \frac{\sigma_{s_c}^2}{\sigma_q^2} \tag{2.34}$$

The quantity $\frac{\sigma_{s_c}^2}{\sigma_q^2}$ can be considered as a *signal-to-noise ratio* for the conversion process. It gives the number of bits required to obtain a fixed quantisation error, when the power of the input signal is known.

2.5.1 THE EFFECTIVE BITS

In a real case a number of sources (inherent in both the A/D converter and the input signal) contribute to the system error. If it is possible to assume that these causes are independent, the cumulative variance can easily be computed and (2.34) can be rewritten as

$$B_e = \frac{1}{2}\log_2 \frac{\sigma_{s_c}^2}{\sigma_c^2} = \frac{1}{2}\log_2 \frac{\sigma_{s_c}^2}{\sigma_c^2}\frac{\sigma_q^2}{\sigma_q^2} = B - \frac{1}{2}\log_2 \frac{\sigma_c^2}{\sigma_q^2} \tag{2.35}$$

where B_e is the *effective bits* of the A/D converter and σ_c^2 is the cumulative noise variance. Equation (2.35) shows that the effective bits is smaller than the bits as expressed by (2.34) and the two quantities are equal only if $\sigma_c^2 = \sigma_q^2$.

It is interesting to study the dependence of the effective bits on the *residual error variance*

$$\sigma_r^2 = \sigma_c^2 - \sigma_q^2 \tag{2.36}$$

Figure 49 shows the dependence of the effective bits as a function of the ratio $\frac{\sigma_{s_c}^2}{\sigma_r^2}$, for various numbers of system bits.

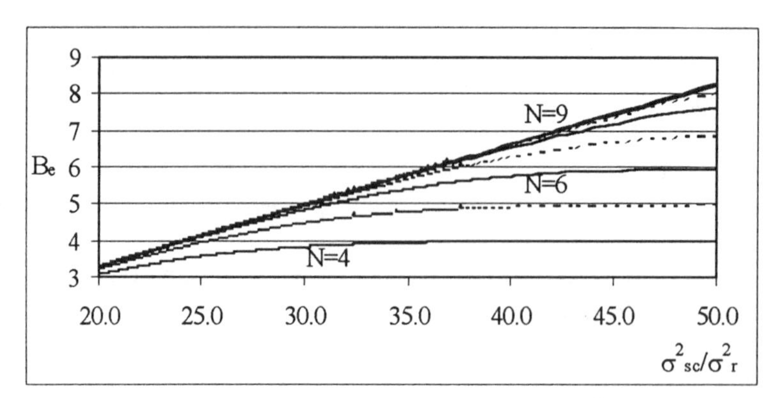

Figure 49. Dependence of the effective bits as a function of the ratio $\frac{\sigma_{s_c}^2}{\sigma_r^2}$, for various numbers of system bits.

Figure 49 shows that when the residual error variance is small, compared to the quantisation error variance, the effective bits becomes the same as the number of A/D converter bits.
When the residual error variance is large the curves merge, independently of the A/D converter bits. Hence, in this case it does not make sense to increase the A/D converter bits and a better solution is to reduce the residual error variance.

2.5.2 OVERSAMPLING A/D CONVERSION

The oversampling technique (i.e. the use of artificially high acquisition rates as compared to the Nyquist rate of the signal) is used to improve the effective bits of A/D converters. This class of converters uses fast sampling clocks and makes extensive use of digital processing, exploiting the suitability of fast digital VLSI circuits. The structure of a typical oversampling A/D converter is shown in Figure 50.
The benefits of the oversampling technique for the quantisation error can easily be understood if the *PSDs* of the processes involved are considered. Let us assume that

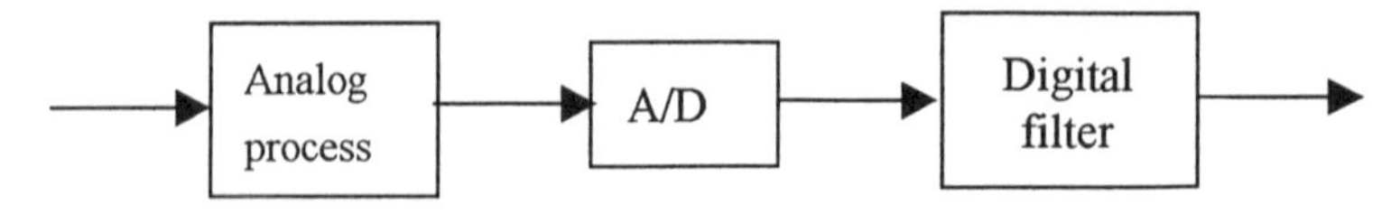

Figure 50. Structure of a typical oversampling A/D converter.

the input signal has a maximum frequency of *0,5F*. The *PSDs* of both the input signal and the sampled signal are shown in Figure 51a and 51b.
Though the corresponding Nyquist frequency is *F*, the sampling rate F_s is adopted

$$F_s = MF \tag{2.37}$$

M being an integer number, called the *oversampling rate*

Equation (2.31) shows that the error variance depends on the quantisation level, hence it does not change if the bits and the input signal amplitude remain constant.
Let us now consider an oversampling A/D converter with the same bits *B* as the Nyquist converter. The total power error for this converter and the Nyquist sampler is the same. Moreover, based on the definition of error variance (see Chapter 1), it is clear that it represents the error power, hence its value also gives the integral of the *PSD*. However, the error power is now uniformly distributed in the frequency interval [*-0,5MF, 0,5MF*], hence the corresponding level is $\frac{\sigma_q^2}{MF}$, as shown in Figure 51c.
The spectrum of the input signal extends to the frequency *0,5F*. It is therefore possible, without any loss of information, to filter the output digital data by using a digital low-pass filter whose cutoff frequency is *0,5F*, as shown in Figure 51d. The error power that is in the bandwidth of interest, i.e. the *baseband error power* is then

$$\sigma'^2_q = \frac{\sigma_q^2}{MF} F = \frac{\sigma_q^2}{M} \tag{2.38}$$

A comparison between the *PSDs* of the quantisation error and the input signal in the case of oversampling is made in Figure 51e.
Clearly the oversampling technique gives a quantisation power error value that is *M* times smaller than the error power given by the corresponding Nyquist converter.
The effective bits increase in accordance with equation (2.35). In fact, considering the noise variance of the oversampling A/D converter as the cumulative error variance (2.35), we get

$$B_e = \frac{1}{2}\log_2 \frac{\sigma_{s_c}^2 M}{\sigma_q^2} = \frac{1}{2}\log_2 \frac{\sigma_{s_c}^2}{\sigma_q^2} + \frac{1}{2}\log_2 M = B + \frac{1}{2}\log_2 M \tag{2.39}$$

It is interesting to determine the relation between the increase of the effective bits of an A/D converter and the corresponding oversampling factor. Equation (2.39) gives

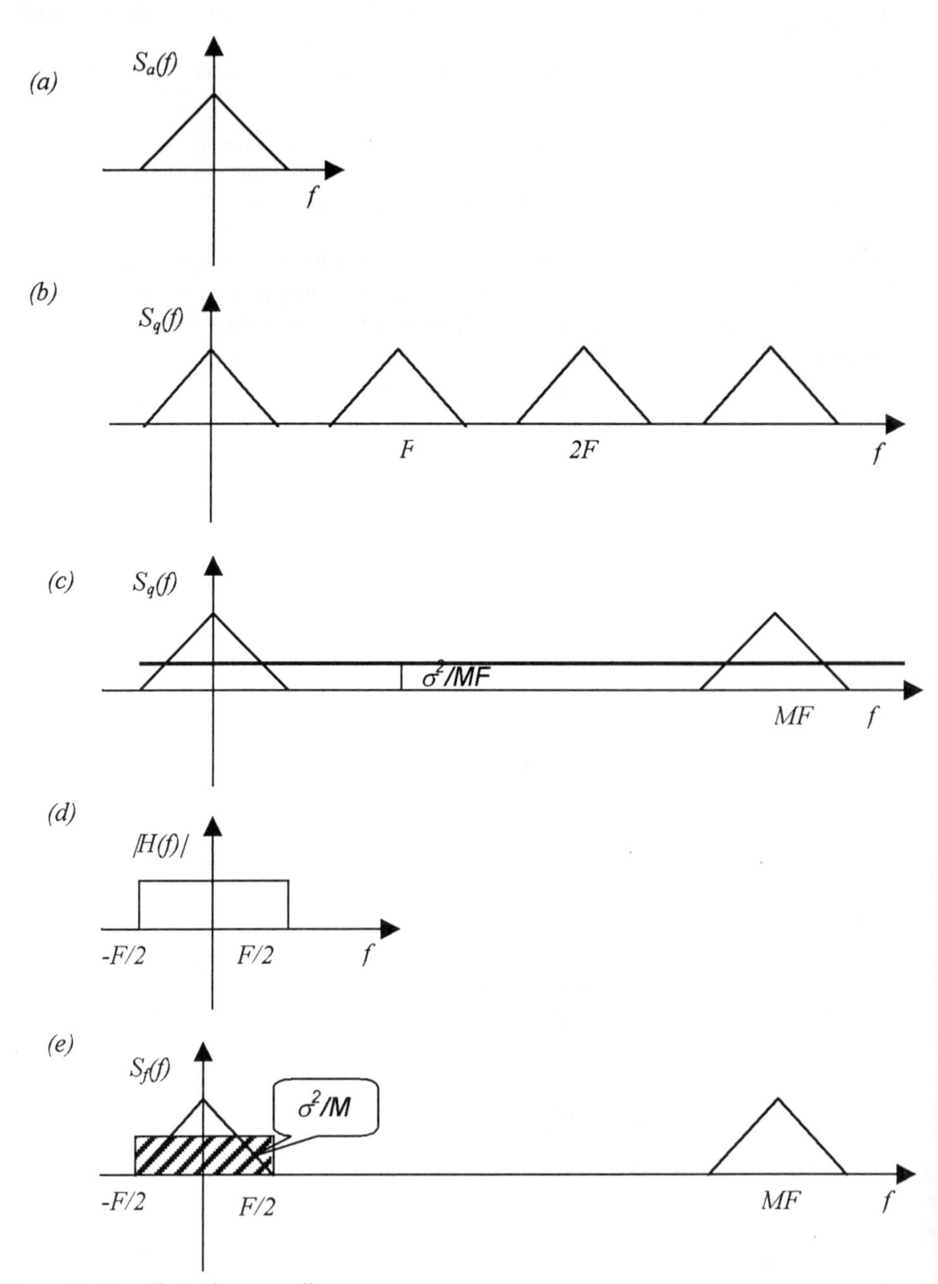

Figure 51. The effects of oversampling.

$$M = 2^{2(B_e - B)} \tag{2.40}$$

Equation (2.40) shows that increasing the effective bits by one requires the oversampling rate to be multiplied by 4. Unfortunately this easily gives oversampling rates that can not be realised with available technologies. Fortunately, oversampling A/D converters are not the most efficient devices to increase the effective bits and a number of techniques have been proposed to obtain better results. These techniques give better results because they produce a *shaping* of the error *PSD* that moves the energy of the quantisation error towards a high frequency. Provided that the analog input to the A/D converter is oversampled, the high frequency quantisation error can be eliminated with a digital low-pass filter. The topic of shaping A/D converters is beyond the scope of this book and interested readers can refer to specific literature.

2.5.3 THE DITHERING TECHNIQUE

Dithering is to the purposeful addition of *low-level* noise to a signal before its quantisation to enhance the resolution of the A/D converter and hence to improve its effective bits B_e. The non-linear nature of quantisation, in fact, always introduces a strongly input-dependent error into the signal. This error can represent a significant signal modification for low-level or simple input signals.
The effects of dithering consist of an improvement of the statistical behaviour of the quantisation error. More specifically, adequately chosen random dither signals can cause the quantisation error to be signal-independent, uniformly distributed, white noise [27]. This means that a significant portion of the error power can be moved outside the signal band and can easily be eliminated by digital low-pass filtering of the quantiser output. In this way the quantiser resolution and linearity are improved at the expense of a reduced conversion rate and input range amplitude [25].
A dithered A/D converter quantisers the signal

$$u_n = s_c + w_n \tag{2.41}$$

w_n being a random process, assumed to be independent of the sampled input signal and is called the *dither* process. The quantiser in Figure 47 is, therefore modified as shown in Figure 52.

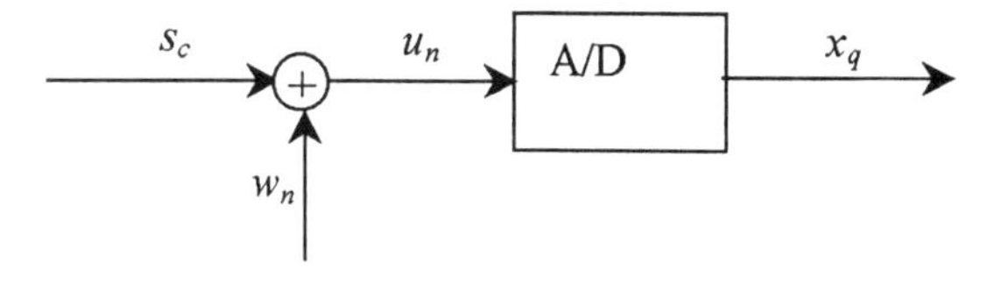

Figure 52. The block diagram of a dithered quantiser.

The idea of dithering is that the addition of the dither forces the quantisation error

$$e_n = q(s_c + w_n) - (s_c + w_n) \tag{2.42}$$

to be independent and identically distributed. It should be observed that the quantity introduced in (2.42) is the error introduced by the quantiser. The corresponding difference between the system input and the final output is

$$\varepsilon_n = q(s_c + w_n) - s_c = e_n + w_n \tag{2.43}$$

and is referred to as the *quantisation noise*.

The dither signal characteristics required in order to guarantee that the quantisation error and the quantisation noise are independent of the input signal, *uniform* and white have been studied and a number of works exist on this topic [25, 27].
For the purpose of this book it will suffice to say that the most well-known case in literature refers to *uniform* dither. It is a known fact that if the dither signal has a uniform *PDF* in the range (-0,5*Q*, 0,5*Q*) then the quantiser error is uniformly distributed in the same interval.
Choice of an adequate dither signal can be much more difficult if the characteristic of the corresponding quantisation noise is of interest. It is not, in fact, obvious in this case that a uniformly distributed dither signal represents the correct choice. A number of works show that choice of the correct dither signal depends, in this case, on the nature of the input signal.
The effects of the introduction of the dithering signal in an AD converter can be evaluated by using the effective bits. According to the previous definition, the effective bits for a dithered quantiser are

$$B_e = B - \frac{1}{2}\log_2\frac{\sigma_d^2}{\sigma_q^2} \tag{2.44}$$

being σ_d^2 the variance of the error in the dithered quantiser. The corresponding increment in the bits obtained by using the dithering approach is

$$B_i = B_e - B = \frac{1}{2}\log_2\frac{\sigma_q^2}{\sigma_d^2} \tag{2.45}$$

In (2.45) the condition $\sigma_c^2 = \sigma_q^2$ is supposed to sake of simplicity.

2.5.4 DITHERING IN THRESHOLD SYSTEMS

The overall effect of dithering consists of linearisation of the system characteristic and it can be successfully applied in cases where linearisation of the characteristic of a threshold system is desired [28]. In a subsequent section of the book dithering will be applied to the linearisation of *quasi-linear* devices, i.e. devices that show a linear input-output relationship, apart from a sub-interval of their working range, where the system presents a threshold behaviour, as shown if Figure 53a. Clearly such a system will not give any output signal as long as the corresponding input signal is smaller than the threshold level Δ. The input and output signals for the system are given in Figure 53b.

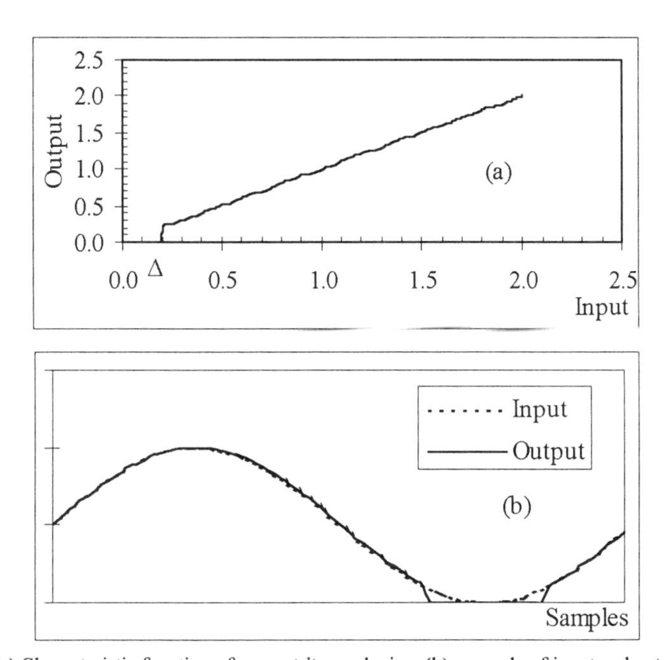

Figure 53. (a) Characteristic function of a *quasi-linear* device; (b) example of input and output signals.

Figure 54 shows an example of the results that can be obtained by using a dithering signal and then filtering the system output. Figure 54a shows the system characteristic obtained by the dithering approach and figure 54b shows the corresponding input-output signals. The figure clearly shows linearisation of the system characteristic.

It is worthwhile pointing out that the capability of linearising the characteristic curve of *quasi-linear* systems is of great importance in the measurement field. A number of measuring devices, in fact, have this type of non-ideal behaviour that limits their application when very low-level signals are considered.

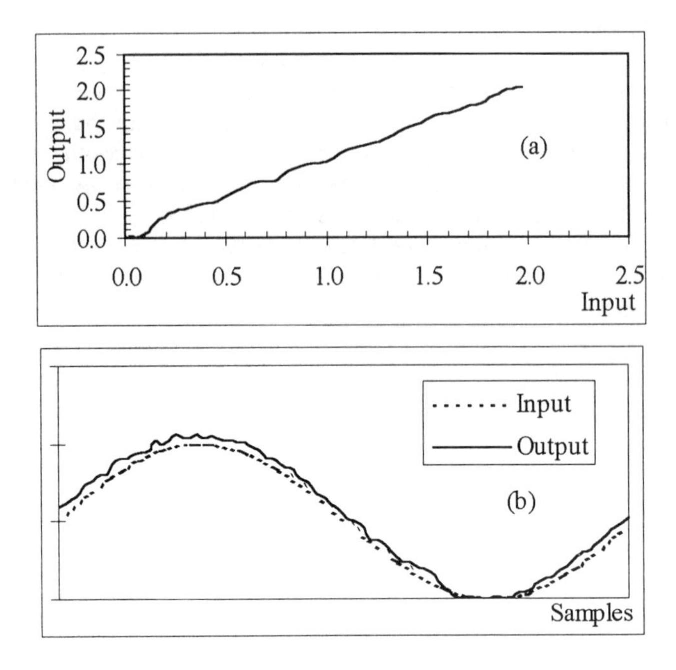

Figure 54. Characteristic function of a dithered *quasi-linear* device (a) and an example of the corresponding input and output signals (b).

A more detailed description of the application of the dithering approach to the linearisation of *quasi-linear* devices will be made later on in the book.

2.6 Stochastic Resonance and dithering

In the light of what has been discussed in the previous sections, we would now like to make a few considerations on the classification of noise-added techniques. This will not be done by recourse to burdensome mathematical justification or enigmatic concepts of abstract physics, but by means of as qualitative engineering-oriented approach, i.e. based on a very specific meaning of the heuristic awareness of the phenomena being dealt with.

How would you answer the question "What is the difference between stochastic resonance and dithering?" After all, they are both basically noise-added techniques, they both aim to enhance the performance of a system by reducing the threshold error, and they both require the optimisation of certain parameters. So why use two names? This is obviously a leading question as bibliographical references have traditionally made a distinction between the two techniques. The idea of adopting a unified approach to determine the optimal working conditions for stochastic systems will be dealt with in greater detail in the following chapters. Here we just want to point out the differences between the two phenomena. One thing is certain: at least genealogically, stochastic resonance is a phenomenon that is observed in the natural world, whereas dithering is an artificial technique used to ensure the applicability of certain assumptions which would otherwise be hard to account for.

In our opinion, a first useful criterion to distinguish between the two noise-added techniques is closely connected with the target to be reached. Stochastic resonance is a condition peculiar to bistable systems, with which it is possible to reduce the system threshold (and thus render the system sensitive to forcing signals with an amplitude lower than the threshold) without altering the system's characteristics (i.e. the systems remains bistable). Dithering, on the other hand, is used to increase the resolution of systems which present sudden non-linearity. This is achieved by linearising the characteristic curve of the system in the proximity of the non-linearity, thus modifying the nature of the system (a system which was bistable, at least in the part affected by non-linearity, is linearised). It is clear that reducing the threshold of a system is quite different from linearising a part of the characteristic.

This initial differentiation enables us to classify quite precisely the effects and fields of application of the two techniques. The basic idea behind stochastic resonance techniques is to determine the level of noise that is capable of maximising the number of interactions between the forcing signal and the system threshold in a known period of time, T, in such a way as to ensure a system transition during the whole of the period T. Evidently, the probability of success depends on the frequency of the forcing signal and the system time constants, if any (i.e. how the system reacts to stochastic forcing).

The idea behind dithering, on the other hand, is to modify the forcing signal with a level of noise that will guarantee interaction between the forcing signal and the system that is proportional to the frequency of the signal. In this case, the weight due to the frequency of the input signal is less, as it is sufficient for the interactions just to occur, however many times. An average operation will then sort it all out (recall that if a bit of a converter operating on 10 samples takes a low value 5 times and a high value 5 times or, when operating on 6 samples takes a low value 3 times and a high value 3 times, the average is always 0.5). Obviously, it is important for the noise bandwidth to be greater than that of the input signal.

It therefore seems clear that the theoretical approaches to the two phenomena will have to be very different. Dithering is applied to systems that can be considered static, at least around the working point. The optimisation formulae do not take the system's time constants into account. Stochastic resonance, on the other hand, is a phenomenon typical of both static systems (e.g. a trigger), where it is important to take the forcing signal frequency into consideration, and dynamic systems (e.g. *Brownian motion*) where the system's time constants also come into play (recall the matching condition, where the characteristic frequencies of the system appear).

Therefore, when the target is different the basic idea, the theoretical approach and obviously the experimental set-up for implementation of the two techniques also differ.
In the next chapter we will use the probabilistic approach to generalise the concept of stochastic modulation.

References

[1] F. Moss and K. Wiesenfeld "The Benefits of Background Noise" Scientific American Aug. 1995, p. 50

[2] R. Benzi, G. Parisi, A. Sutera and A. Vulpiani "Stochastic resonance in climatic change" Tellus, 34 10 (1982)

[3] C.Nicolis "Stochastic aspects of climatic transitions - response to a periodic forcing" Tellus 34 1 (1982)

[4] R. Benzi, G. Parisi, A. Sutera and A. Vulpiani "A theory of stochastic resonance in climatic change" SIAM J. Appl Math 43, 565 (1983)

[5] McNamara K. Wiesenfeld and R. Roy "Observation of Stochastic Resonance in a ring laser" Phys. Rev. Lett. 60 2626 (1988)

[6] P. Jung and P. Hanggi "Stochastic nonlinear dynamics modulated by external periodic forces" Europhys. Lett. 8 505 (1989)

[7] G. Matteucci "Orbital forcing in a stochastic resonance model of the late Pleistocene climatic variations " Climate Dyn. 3, 179 (1989)

[8] L. Gammaitoni N. Marchesoni, E. Menichella-Saetta and S. Santucci "Stochastic resonance in a bistable systems" Phys. Rev. Lett. 62 349 (1988, 1989)

[9] N. Marchesoni, E. Menichella-Saetta, M. Pochini and S. Santucci "Analog simulation of underdamped stochastic systems driven by colored noise:spectral densities", Phys. Rev. A 37, 8, 1988.

[10] M. I. Dykman, A.L. Velikovich, G.P. Golubev, D.G. Luchinskii and S.V. Tsuprikov "Stochastic resonance in an all-optical passive bistable system" Soviet Phys JETP Lett 53, 193 (1991)

[11] R.N. Mantegna, B. Spagnolo "Stochastic resonance in a tunnel diode" Phys. Rev. E49 R1792 (1994)

[12] R. Benzi, A. Sutera and A. Vulpiani "The mechanism of Stochastic Resonance" J. Phys. A: Math. Gen.14L 453 (1981)

[13] J.P. Eckmann and L.E. Thomas "Remarks on stochastic resonance" J. Phys. A: Math. Gen. 15 L261 (1982)

[14] S. Fauve and F. Heslot, "Stochastic resonance in a bistable system" Phys. Lett. 97A 5 (1983)

[15] B. McNamara and K. Wiesenfeld "Theory of Stochastic Resonance" Phys. Rev. A39 (1989)

[16] T.L. Carrol and L.M. Pecora "Stochastic Resonance and Crises" Phys. Rev. Lett. 70 576 (1993)

[17] M.I. Dykman, R. Mannella, P.V.E. McClintock and N.G. Stocks "Comment on Stochastic Resonance in bistable systems" Phys. Rev. Lett. 65, 2606 (1990)

[18] L. Gammaitoni, F. Marchesoni, E. Menichella-Saetta and S. Santucci "Reply to comment on: Stochastic Resonance in Bistable Systems" Phys. Rev. Lett. 65, 2607 (1990)

[19] L. Gammaitoni "Stochastic resonance in multi-threshold systems" Phys. Lett. A 208 315 (1995)

[20] Gammaitoni L. Hanggi P. Jung P. Marchesoni F. "Stochastic Resonance" Reviews of Modern Physics. 70(1):223-287, 1998 Jan.

[21] Dykman MI. Mcclintock PVE. "What can stochastic resonance do" Nature. 391(6665):344, 1998 Jan 22.

[22] M.F. Wadgy, "Effect of various dither forms on quantizzation errors of ideal A/D converters", IEEE Trans. Instrum. Measu., vol.38, n.4, 1989.

[23] L. Schuchman, "Dither signals and their effect on quantizzation noise", IEEE Trans. Commun. Technol., vol.12, 1964.

[24] J. Vanderkooy and S.P. Lipshitz, "Dither in digital audio", J. Audio Eng. Soc., vol.35, n.12, 1987.

[25] P. Carbone, D. Petri, "Effect of Additive Dither on the Resolution of Ideal Quantisers",IEEE Trans. Instrum. Meas.; 1994, 43,3, 389-396

[26] Gammaitoni L., Marchesoni F., Santucci S., "Stochastic Resonanc as a Bona Fide Resonance", *Phys. Rev. 74/7, 1995, 1052.*

[27] R. M. Gray, T. G. Stockoham, Dithered Quantisers, IEEE Trans. Inf. Th..; 1993, 39,3, 805-812

[28] L. Gammaitoni, "Stochastic resonance and the dithering effect in threshold physical systems", Phys. Rev. E.; 1995, 52,5, 4691-4698

[29] A. Papoulis, "*Probability, Random Variables, and Stochastic Process*", McGRAW-HILL BOOK COMPANY.

[30] J.M.T. Thompson and H.B. Stewart, "*Nonlinear dynamics and chaos*", John Wiley and Sons, 1989.

3 A PROBABILISTIC APPROACH TO NOISE-ADDED SYSTEMS

3.1 Introduction

The main features of noise-added systems can be summarised as the possibility of both improving the system's features without destructive action on the device and optimising the performance of systems operating in a noisy environment [1-9].
When either a linear or a bistable system is forced with a noise signal both its Probability Density Function (*PDF*) and the optimal value of its standard deviation σ_{opt} must be determined, in order to characterise the interaction between the noise and the system.
Several studies have been devoted to the choice of a suitable form for the *PDF* to allow the imposed target to be reached. However, very often the main problem in dealing with noise-added systems remains the identification of an optimal value for the noise standard deviation σ_{opt}. Generally, the σ_{opt} value is obtained by maximisation of the Signal-to-Noise Ratio [1-3]. The drawbacks related to this approach can be summarised as follows:

- the performance required of the system cannot be assured. For example, the duty-cycle of the output signal in a bistable system can be ruled by fixing a suitable noise amplitude;

- constraints on the system behaviour (e.g. output roughness) cannot be taken into account. This mainly is a problem with measurement devices. In fact noise-added techniques are often not considered because of the impossibility of dirtying a system by forcing an external noise signal, even if good performance could be obtained. The possibility of applying noise modulation techniques to reduce the physical threshold of some electronic devices (comparators, sensors) is reliable but at the same time a minimum noise level is required;

- when system parameter fluctuations occur, analog implementation of an adapting system allowing noise amplitude tuning, based on calculation of the signal-to-noise ratio, is very complex and real-time control of the variance value is not allowed;

- the influence of external parameters on system performance cannot be taken into account. This task becomes a duty especially for bistable systems which are characterised by a strict dependence of the optimal noise amplitude on the system parameters [10].

As a matter of fact the *matching condition* can sometimes be used to determine the optimal noise standard deviation, σ_{opt}, in bistable systems [10]. It states that

$$T=2\,T_K \tag{3.1}$$

where T is the forcing signal period and $R_K=1/T_K$ is the well-known *Kramers Rate* [10]. For example, if a double well potential subjected to both dissipation and fluctuation is considered, the *Kramers Rate* assumes the following form

$$R_K = \frac{\omega_0 \omega_b}{2\pi\gamma} e^{-\frac{DV}{D}} \tag{3.2}$$

ω_0 and ω_b being the characteristic frequencies of the potential, DV the system threshold and D a quantity proportional to the noise standard deviation σ [10].

However condition (3.1) cannot be used to determine the σ_{opt} value for any class of noise-added systems.
To counter these drawbacks an approach to optimal noise variance detection, based on the optimisation of a more general index than the signal-to-noise ratio, can be used [11]. The index to be optimised should be defined by respecting the required system performance, (e.g. system switching at the forcing signal frequency), and accounting for the system constraint (e.g. output roughness).
In further sections the possibility of improving both bistable and linear measurement device performance will be investigated [12-15].

3.2 An overview of the optimisation approach

A schematic representation of this general approach is given in Figure 1. The main idea is to define, for each class of systems, a functional W to be optimised, the signal-to-noise ratio being but one of the options. Dealing with stochastic systems, the index W would very often be based on a probabilistic approach. Analytical optimisation of the index W will lead to the general relationship between the system parameters and the optimal standard deviation σ_{opt} [11]

$$\sigma_{opt}=f(S,\ \nu,\ A,\ \rho,\ F(N)) \tag{3.3}$$

where

σ_{opt} is the optimal noise variance,
S is the system threshold,
ν is the forcing signal frequency,
A is the forcing signal amplitude,
ρ is a parameter allowing the desired system performance to be fixed,
$F(N)$ is a function to be fixed linking the theoretical law to the physical device,
N is a parameter taking into account the influence of the limited band noise.

The *PDF* of the noise signal (*Gaussian*, *uniform*, etc.) to be used depends on the specific application.
When a suitable form of (3.3) has been obtained, a map giving the σ_{opt} value as a function of the system parameter functions is obtained. The map is generated by an algorithm that screens the parameters A and ν and obtains the corresponding σ_{opt} values, by means of (3.3), and must be adopted to optimise the noise features on the basis of the system state. A suitable procedure fixing the $F(N)$ quantity in (3.3) and allowing the physical system to be matched to the theoretical control law also has to be developed.
How it is possible to define the expression $F(N)$ by a minimum set of experimental measures will be discussed in a following section. It will be emphasised that this feature is of great importance to avoid the time-consuming task of obtaining a set of measures mapping the system parameter space $(S, A, \nu)_i$ to the optimal noise variance values σ^i_{opt}, i being the measurements index.
Sometimes the value of the system threshold cannot be directly included in (3.3), being not comparable to the σ value. However, in the case of electrical systems, for example, the relationship between the system threshold S' and the corresponding electrical threshold S can easily be detected by system characterisation (i.e. in IR sensors the physical threshold is represented by the maximum distance S' while the electrical threshold S is the minimum value of the signal correctly detected by the receiver). The relationship between these two values derives from the system characterisation

$$S=g(S') \tag{3.4}$$

where $g(\bullet)$ represents the analytical expression of the system characteristic and must be included in expression (3.3).

If the noise signal amplitude value is not comparable to the system threshold, experimental trials allowing collection of the data set $(A,\ \nu,\ \sigma_{opt})$, to be used for generation of the *iso-amplitude curves*, are unavoidable.

In addition, before processing relation (3.3) for the map generation, a test on the accuracy of the form identified for the relation must be performed.

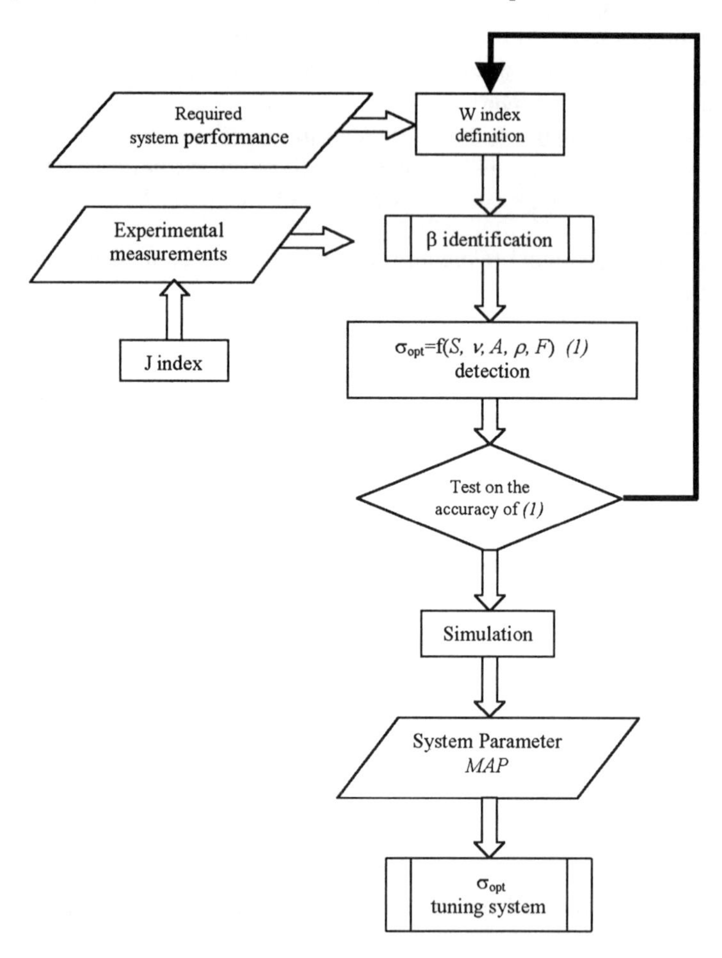

Figure 1. A schematic representation of the proposed approach. It is valid for any class of systems and allows for the analytical identification of the relationship between optimal noise variance and system parameters.

3.3 A form for the index *W* in a bistable system

In this section the *SR* phenomenon, based on the non-deterministic signal theory applied to stochastic systems, is considered. A periodical signal, modulating the threshold S of a bistable system, and a *Gaussian* noise signal $N(t)$, with zero mean and σ^2 variance, are shown in Figure 2a. As can be observed, the interaction between the two signals allows for system commutations during a short time interval, each time the periodical signal takes the minimum value, as is shown in Figure 2b. In order to avoid unsuitable random fluctuations of the system output, being superposed over the periodical commutation, the noise amplitude must be contained inside a suitable interval $I=[\rho, S]$. Noise fluctuations greater than S, producing system commutations, would make it impossible to reconstruct correctly the information linked to the input signal. This kind of occurrence is shown in Figure 2. The quantity marked ρ in Figure 2 has been called *commutation depth* and it defines the amplitude of the periodic signal involved in noise interactions. Although it can be stated that the higher ρ is the longer the commutation period is, too high a value of ρ causes a rise in the noisy behaviour of the system. Suitable bounds for the values allowed for ρ must therefore be defined. The lower bound of ρ depends on the periodical signal amplitude A. In fact, in order to assure signal interactions, it must be

$$\rho \geq S-A \tag{3.5}$$

while the upper bound depends on the system performance.
Actually, the value of the commutation depth on the system parameters is

$$\rho=\varphi(S-A) \tag{3.6}$$

where $\varphi(.)$ is a general function to be defined on the basis of the required system performance.

For example, in order to investigate the minimum value of the forcing amplitude, A_{min}, allowing system commutations, equation (3.6) becomes

$$\rho=S-A_{min} \tag{3.7}$$

In order to define a functional W, called the commutation probability, that represents the probability that system commutations will occur with the same frequency as the forcing periodical signal, some considerations must be made.
In further discussions, interesting consideration about the dependence of W on the commutation depth ρ, allowing for the optimisation of the system performance, will be made.

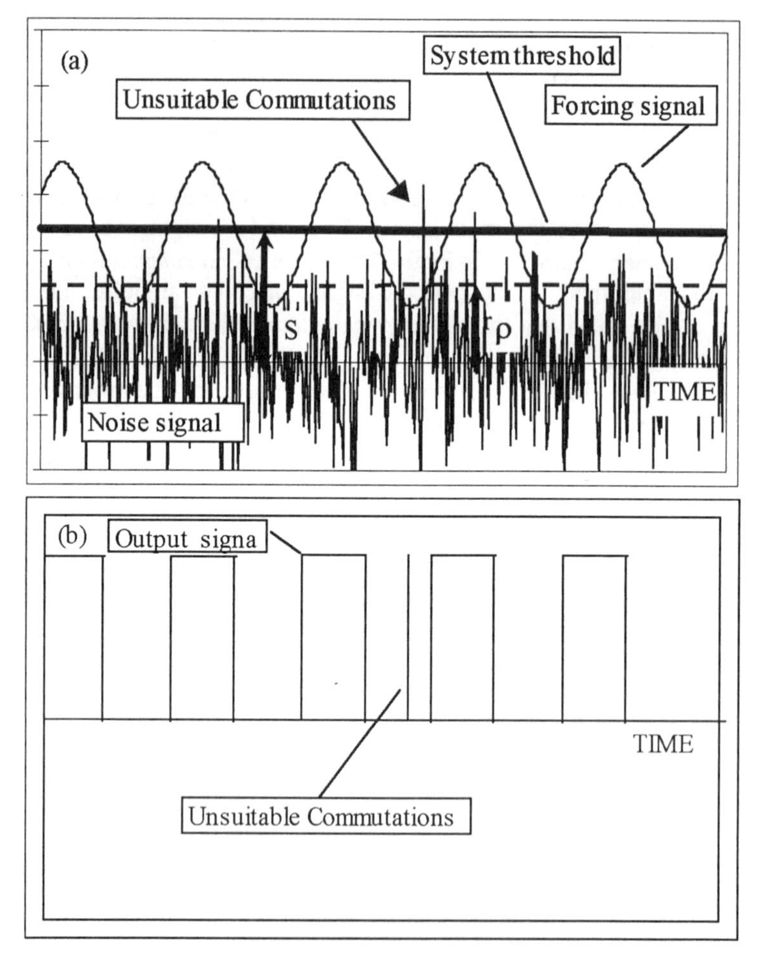

Figure 2. (a) Periodical signal, modulating the threshold S of a bistable system, and *Gaussian* noise signal $N(t)$, with zero mean and σ^2 variance; (b) plot of the system commutation.

3.3.1 THE MATHEMATICAL DETERMINATION OF OPTIMAL NOISE VARIANCE

The probability of success, p, that noise fluctuations will be inside the interval I allows us to detect whether the input signal is transferred to the system output or not. The form of this probability is the following

$$p=P(\rho<N(t)<S) \qquad (3.8)$$

The quantity p states the possibility of forcing a threshold system by an under-threshold signal. In general, in order to investigate the probability that a random variable is outside an arbitrary interval $[-D,D]$ the following Tchebycheff inequality can be used [16]

$$P(|x| > D) < \frac{\sigma^2}{D^2} \tag{3.9}$$

For specific probability densities, the bound introduced by the Tchebycheff inequality is too high. Better results can be achieved using the following approach.
Let us assume that a random variable y is normally distributed, its probability density function is

$$g(y) = \frac{1}{\sqrt{2\pi}} e^{-\frac{y^2}{2}} \tag{3.10}$$

and its cumulative distribution function is given by

$$G(x) = \int_{-\infty}^{x} g(y)dy \tag{3.11}$$

$G(x)$ can also be expressed in terms of the error function

$$erf(x) = \frac{1}{\sqrt{2\pi}} \int_{0}^{x} e^{\frac{-y^2}{2}} dy = G(x) - \frac{1}{2} \tag{3.12}$$

If the considered distribution has a variance equal to σ, equations (3.10) and (3.11) must be scaled

$$f(x) = \frac{1}{\sigma} g(x/\sigma) = \frac{1}{\sigma\sqrt{2\pi}} e^{\frac{-x^2}{2\sigma^2}} \tag{3.13}$$

$$F(x) = G(x/\sigma) \tag{3.14}$$

The probability that a random variable x is smaller than D can be evaluated by

$$P(x<D) = F(D) = G(D/\sigma) = erf(D/\sigma) + 0.5 \tag{3.15}$$

Hence the probability $p = P(\rho < N(t) < S)$ is

$$p=P(\rho<N(t)<S)=P(N(t)<S)-P(N(t)<\rho)= \\ =G(S/\sigma)-G(\rho/\sigma)=erf(S/\sigma)-erf(\rho/\sigma) \tag{3.16}$$

Figure 3 shows the probability p as a function of both the noise variance σ and the commutation depth ρ. The dependence of p on σ assures the possibility of controlling the system performance by acting on the noise level. It should be underlined that the value of ρ should be kept as low as possible, respecting (3.5), if a low level of noise is required (system constraint). On the other hand large values of ρ could be required for better system behaviour (system performance). Hence it can be stated that the required system performance and the system constraints define the upper bound for ρ. Moreover, it can be observed that, as expected, the noise variance assuring a suitable value for the probability P increases with ρ.

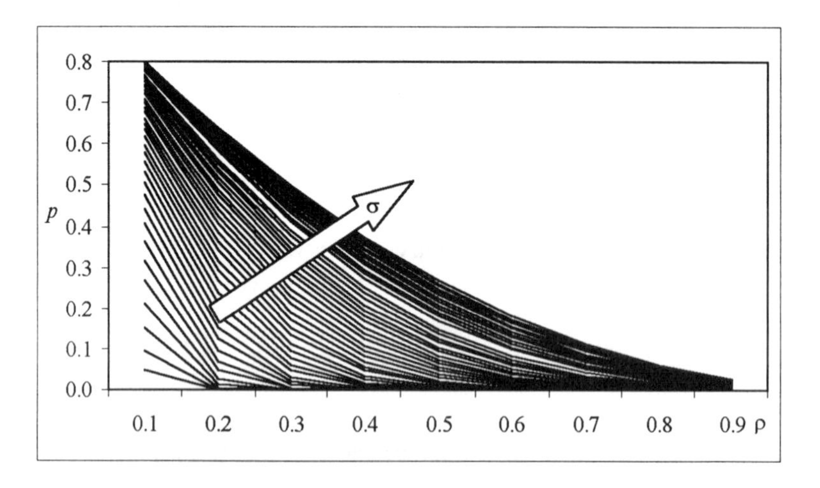

Figure 3. Probability p as a function of both the noise variance σ and the commutation depth ρ.

Now the interaction between the noise signal and the periodical signal will be investigated. If N is a parameter linked to the noise spectrum width amplitude and K is a parameter related to the periodic signal frequency, the probability W that K times of N the noise crosses the input signal is given by the *Bernoulli theorem* [16]

$$W_{K,N}=\begin{bmatrix} N \\ K \end{bmatrix} p^{K} q^{N\quad K} \tag{3.17}$$

where

- p corresponds to the success probability introduced above;
- $q=1-p$.

To add a degree of freedom to the definition of the operational typology of systems belonging to the class being investigated (trigger, comparators, etc.) it is necessary to introduce another parameter, M. Parameter M is given the name of *transition factor* and represents the percentage of transitions necessary to meet specifications. The relation between this parameter and the previous ones is as follows

$$K=M*K_f \tag{3.18}$$

where
K_f represents the forcing signal frequency parameter.

Figure 4 shows the probability W as a function of the noise standard deviation σ and the signal frequency ν. It can be observed that the behaviour of W is quite similar to the signal-to-noise ratio normally referred in *SR* theory. When the noise standard deviation matches the optimal value, allowing for W to be optimised, the required *threshold reduction* is achieved.
The dependence of W on the success probability p and K is shown in Figure 5. It can be proved that the value of p assuring a maximum value for the probabilities W is equal to K/N.

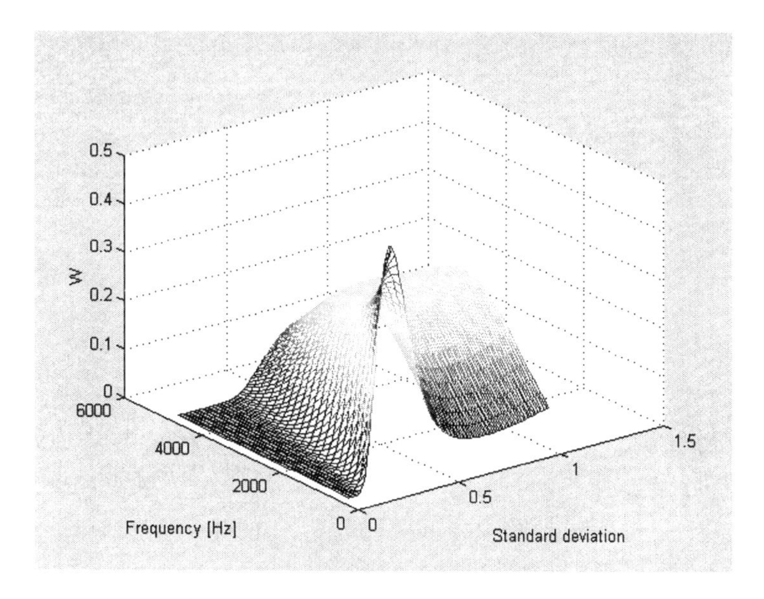

Figure 4. Probability W as a function of σ and the forcing signal frequency ν.

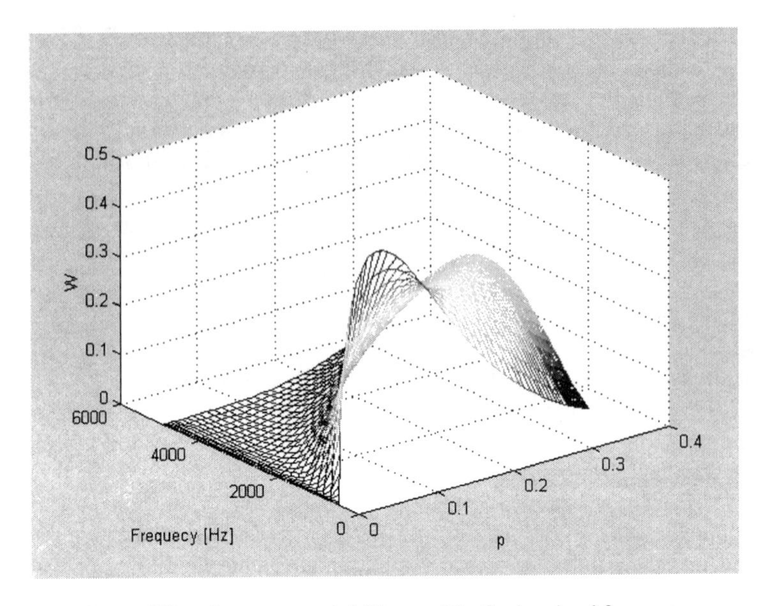

Figure 5. Dependence of W on the success probability p and the forcing signal frequency ν.

$$W = \binom{N}{K} p^K (1-p)^{N-K}$$

$$\frac{dW}{dp} = \binom{N}{K}\left[Kp^{K-1}(1-p)^{N-K} - (N-K)p^K(1-p)^{N-K-1}\right]$$

$$\frac{dW}{dp} = 0 \qquad \Rightarrow Kp^{K-1}(1-p)^{N-K} = (N-K)p^K(1-p)^{N-K-1} \tag{3.19}$$

$$\frac{K}{N-K}p^{-1}(1-p) = 1 \qquad \Rightarrow p\left(1+\frac{K}{N-K}\right) = \frac{K}{N-K}$$

$$\Rightarrow p = \frac{K}{N}$$

This means that, for a fixed K value, the optimal variance assuring a maximum value for W is the same variance assuring the value K/N for p

$$p(\sigma_{opt}) = K/N \tag{3.20}$$

It can be stated that the optimal value of p guaranteeing a maximum value for W changes as K increases. In this case, in order to match the optimal condition again a new value of p must be sought. As can be observed in Figure 3, this condition can be reached by varying the σ value.
This result, consisting of the dependence of the optimal noise value on the forcing frequency, is very important and its correspondence with the experimental trials will be addressed later on. By using equations (3.16) and (3.20) it follows that

$$K/N=erf(S/\sigma_{opt})-erf(\rho/\sigma_{opt}) \quad (3.21)$$

This represents the noise level control law to be searched for, allowing for the optimal value for one system parameter by fixing the remaining parameters.
For example, by fixing the noise variance σ and the forcing frequency ν, a suitable forcing amplitude value A (ρ=S-A) can be calculated or, by fixing ν and A, a suitable optimal noise variance can be calculated. Moreover, if the dependence of the optimal noise variance on the forcing frequency can be mathematically identified, full analytical parameter control can be implemented.

3.3.2 MATCHING THE PHYSICAL SYSTEM: EVALUATION OF THE *F(N)* EXPRESSION

On the basis of the considerations made in the previous sections, some theoretical results can be given in order to match some system parameters with the input signal characteristics.
Assume that the minimum forcing signal amplitude, A_{min}, allowing system commutations and the corresponding minimum noise variance, σ_{opt}, have been detected for discrete values of the forcing frequency [11]. Hence, the experimental pairs $(A_{min}, \sigma_{opt})^i$ have been identified, i=,$1...,K$ being an index numbering the sampled steps of the frequency range considered.
In order to chose the pairs an experimental index can be used. The analytical form of this index could be the following

$$Y=\sum_{i\to 1}^{N}\left|f^i_{csim}-f^i_c\right| \quad (3.22)$$

where

f_{csim} and f_c are the commutation frequency obtained by simulations and the required commutation frequency computed for each period;
N represents the number of periods analysed.

This index has to be minimised and the minimum noise variance, allowing for system commutation with the minimum signal amplitude represents the optimal noise variance σ_{opt}. Hence, the pairs (A_{min}, σ_{opt}) have been identified for discretised values of the forcing frequency.
As an example, Figure 6 shows the simulated functional index J as a function of the noise standard deviation and forcing amplitude, for one forcing frequency value.
In order to find an optimal value for the function $F(N)$, assuring the link between the physical system and the theoretical law, its possible values are screened. For each trial, by using Equation (3.21) the noise variance σ_{th} corresponding to A_{min} is determined

$$A_{min}=S-\sigma_{opt}\ inverf\,[erf(S/\sigma_{opt})-F(N)] \tag{3.23}$$

where *inverf(.)* is the inverse of the error function.
Let us assume that M different values for $F(N)$ have been considered; the described procedure will give M x K pairs $(A_{min}, \sigma_{th})^{i,q}$, $q=1,\dots,M$ $i=1,\dots,K$.

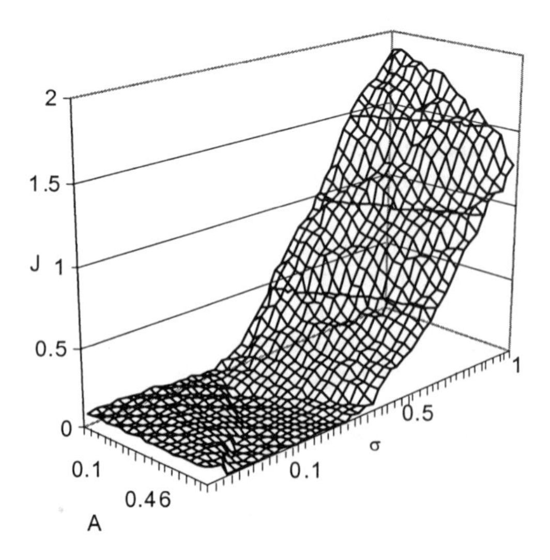

Figure 6. Simulated functional index J as a function of the noise variance and forcing amplitude.

In order to choose the optimal $F(N)$ value among the M investigated, a performance index J_q is required. For example, for each value of $F(N)$ the mean square error between the K experimental values of the noise variance σ_{opt} and the theoretical one σ_{th} could be computed

$$J_q = \frac{1}{K}\sum_{i=1}^{K}\left(\sigma_{opt}^{i} - \sigma_{th}^{i,q}\right)^2, \; q=1...M. \tag{3.24}$$

Hence the value sought for *F(N)* is given by the following expression

$$\mathcal{F}(\mathcal{N}) = \mathcal{F}_{\bar{q}}(\mathcal{N}) \text{ with } \bar{q}: \; min(J_q) = J_{\bar{q}} \tag{3.25}$$

The procedure illustrated above, allowing detection of the optimal value for the expression *F(N)* and assuring that the theoretical law fits the physical device, is sketched in Figure 7.
In order to show the suitability of the proposed approach, some simulations were performed and the simulated results were compared with the theoretical results.
A simulation of a bistable system with a threshold S and hysteresis forced by a periodical signal and a *Gaussian* white noise is now described.
In this case, on the basis of the required system performance, the following dependence of the commutation depth ρ on the threshold S and the forcing amplitude A has been assumed: $\rho=S-A_{min}$.
A comparison between the simulated value of A_{min} and its theoretical values, obtained by using the F function identified in equation (3.23), is made in Figure 8a.
Figure 8b shows the comparison between the *simulated value of* σ_{opt}^{i} *and* $\sigma_{th}^{i,\bar{q}}$, obtained with similar consideration. As can be observed, a good match between the experimental and the theoretical noise variance is obtained.
After *F(N)* has been defined, the whole parameter map can be estimated, representing the first step towards performing optimal noise variance control.
Figure 9 shows the *iso-amplitude* curves as a function of both the noise variance and the forcing frequency, computed by using equation (3.23). By using this map and fixing two of the three parameters, the optimal value of the remaining one allowing for system commutations can easily be computed.
The possibility of performing F(N) detection via a small number of measurements and then estimating the whole iso-amplitude map by a numerical procedure represents the main feature of the proposed approach.
As an alternative solution, the whole parameter map should be experimentally identified, but a larger number of measurements would be required.
As can be observed in Figure 9, and as expected, at a fixed frequency the noise intensity decreases with A. By means of this consideration the noise intensity can be reduced on the basis of the forcing signal amplitude value. Hence, in the study case being considered, the proposed approach allows for minimisation of the forcing noise amplitude.

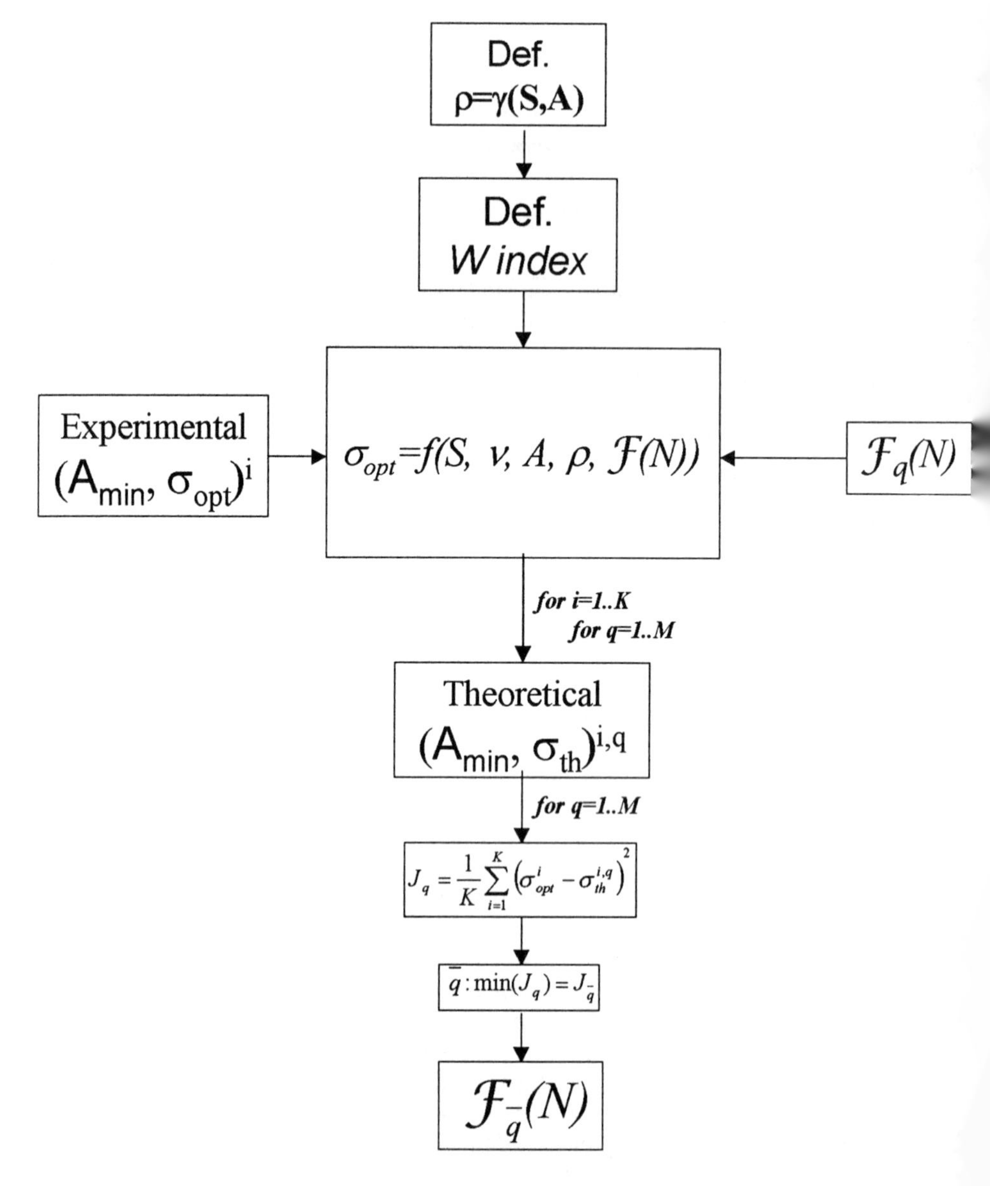

Figure 7. Procedure allowing detection of the optimal value for the expression *F(N)* and assuring that the theoretical law fits the physical device.

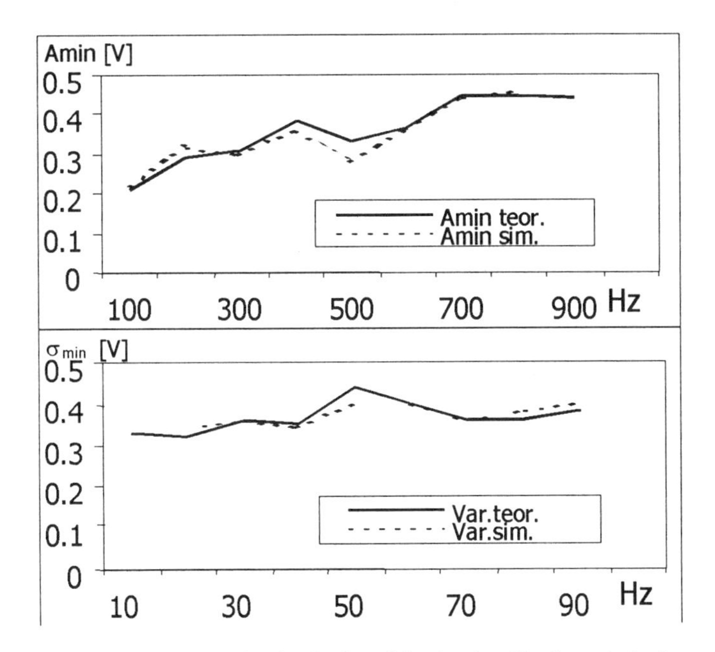

Figure 8. A comparison between the simulated value of A_{min} (σ_{min}) and its theoretical values, obtained by using the identified F function in equation (3.22).

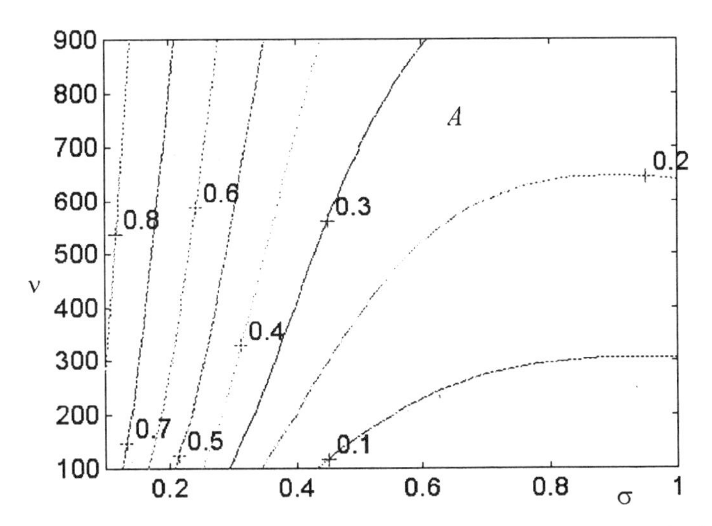

Figure 9. *Iso-amplitude* curves as a function of both the noise variance and the forcing frequency, computed by using equation (3.23).

Anyway, equations (3.6) and (3.21) do not represent sufficient conditions for system performance improvement. For example, if the forcing frequency and amplitude are fixed, the optimal noise variance calculated by using equation (3.21) leads to a system performance improvement only if $A>A_{min}$. This means that, the identification of the pairs $(A_{min}, \sigma_{opt})_K$, for optimal parameter control, represents a fundamental step in the optimisation procedure. On the basis of the above considerations, the *Stochastic Resonance* phenomenon can be characterised as follows:

In general it can be stated that, for a system with a threshold S, the SR effect is a particular working state in which co-operation between a noise signal and a periodical signal, with an amplitude lower than S, allows an improvement in system performance.

In order to characterise the effects and features of a noise-added system it is necessary to optimise an index $W=f(S,\sigma,A)$, joined with the commutation probability. Its dependence on the noise variance allows for detection of optimal system working conditions.

The index W can be defined on the basis of both the particular system and the required target. Moreover, the suitability of the applied method must be confirmed by investigating the improvement in the performance of the system being considered. For example, in literature [1-3], the *SR* condition in bistable systems has been investigated by observing the dependence of R on σ, while improved performance of the system is achieved by threshold reduction.

Hence, it can be stated that the second goal reached with the proposed approach is the possibility of generalising the concept of SR.

The theory introduced above can also be applied when a noise signal $N(t)$ is added to the periodical forcing signal $x(t)$, as is shown in Figure 10. In fact, in this case the success probability p becomes

$$p=P(x(t)+N(t)>S)-P(N(t)>S)= P(N(t)>S-x(t))-P(N(t)>S) \qquad (3.26)$$

and with small forcing signal amplitude values the same condition as equation (3.16) is reached.

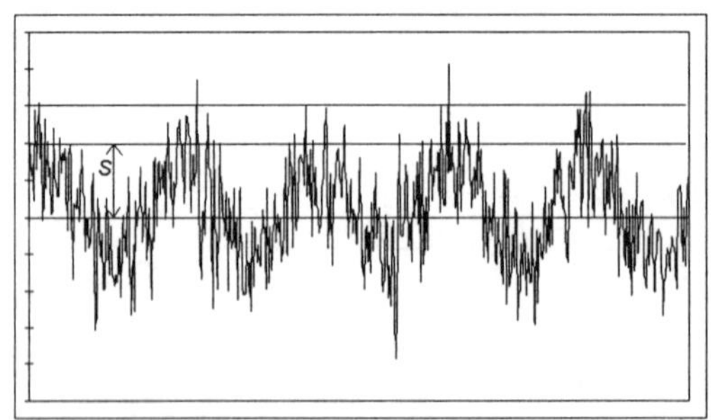

Figure 10. The theory introduced above can also be applied when a noise signal $N(t)$ is added to the periodical forcing signal $x(t)$.

3.4 Noise modulation in quasi-linear systems

The study of noise-added systems also covers quasi-linear systems, i.e. linear systems that in reality present inherent non-linearity in a part of their working range. As with bistable systems, which were discussed in detail on the previous section, the aim of investigating this class of systems is to improve their performance. The possibility of achieving this aim is strictly linked to that of linearising their characteristics in areas affected by saturation or threshold phenomena. Figure 11 is a scheme of the real characteristics of a system of the type being considered and the characteristics to be achieved by applying stochastic modulation techniques.
Observe that, unlike bistable systems, an improvement in the performance of quasi-linear systems is achieved by modifying the shape of the input-output. In systems with two or more stable states, in fact, the threshold was reduced but the topology of the characteristic itself was left unaltered. In linear systems, on the other hand, action on the non-linear areas involves changing the type of the original system.

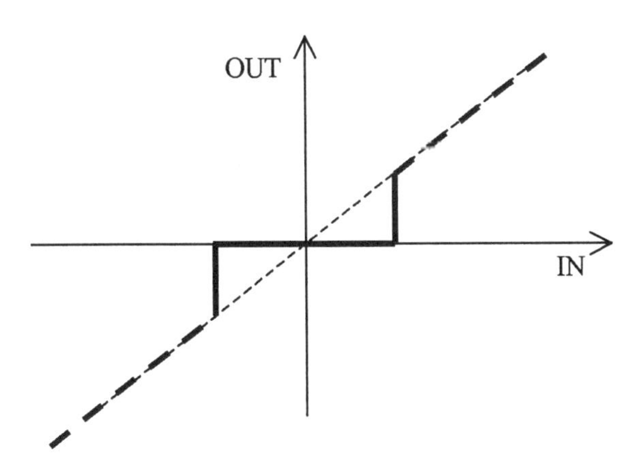

Figure 11. A system with a linear I/O presents a threshold error in the proximity of null input signal amplitudes. This situation is represented by the line in bold type. The thinner line represents the ideal system characteristic.

3.4.1 AN OVERVIEW OF QUASI-LINEAR SYSTEM

One of the most widespread phenomena based on the effects of stochastic modulation is dithering. The theory behind the dithering phenomenon was discussed in detail in the second part of Chapter 2.
Bearing in mind the scale-based feature of a converter and the link between the quantisation error and the amplitude of the levels of quantisation, it can be concluded that reduction of the quantisation error is equivalent to linearisation of the

system characteristic. This is typical of applications based on what is called the linearising effect of *dithering*.
In the field of measurement and linear sensors affected by non-linearity known as threshold error, techniques of this kind can be used to improve the performance of measurement devices [17, 18].
The method proposed is based on the concept that a linear system with threshold non-linearity can be treated, in that working area, as a bistable system whose characteristic needs to be linearised. This situation is illustrated in Figure 12.
From now on we will extend the considerations made on dithering, as a phenomenon closely connected with A/D conversion systems, to pseduo-linear stochastic systems.
The dithering technique has a post-processing stage in which the device's output signal is processed. This stage, which involves calculation of the average of quantised samples, by means of which the system is linearised, is an integral part of the technique being discussed. Consequently, to improve the characteristics of this kind of system it will be necessary to use more complex signal conditioning sections than those used in bistable systems. At the same time, however, the system will be found to be less sensitive to variations in the forcing signal parameters; dither signal modulation systems are therefore less complex.
It is recalled that added-noise systems require a method to calculate the noise variance that will allow them to work in optimal conditions. As was done for bistable systems, we will therefore develop an approach for quasi-linear systems that determines the link between optimal stochastic forcing signal parameters and the physical parameters of the system.

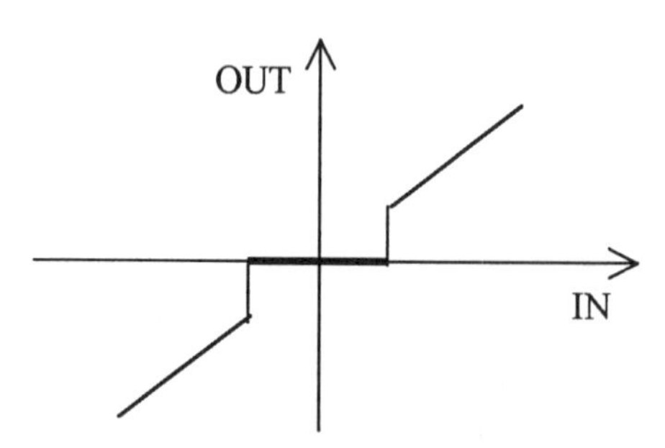

Figure 12. A linear system with threshold non-linearity can be treated, in that working area, as a bistable system whose characteristic needs to be linearised.

3.4.2 THRESHOLD ERROR REDUCTION IN MEASURING DEVICES

Before tackling the problem of determining optimal working conditions for systems with dither, we will illustrate some possible applications of the method proposed.
As observed in previous chapters and further highlighted in the first sections of this chapter, non-linearity in pseudo-linear systems can be eliminated by forcing the system with stochastic components and using suitable techniques for the post-processing of the output signal.
Although the examples given below are only qualitative in nature, they are of use to stress the importance of the problems being discussed and the techniques proposed. In Part 2 of this volume we will present some experimental applications to improve the performance of typical measurement devices. Of course the experimental approach will introduce a certain degree of complexity in both the procedure and the circuit. We have therefore decided to give some examples which although descriptive, have the advantage of focusing on the idea on which the proposed techniques are based.
Let us consider a system like the one shown in Figure 13. The system comprises a pair of optical sensors for the remote transfer of information and it features a threshold error that prevents the transmission of signals whose amplitude is lower than a certain value, *S*. This error is typical of devices used for transmission and reception such as diodes or transistors, which are often used to control areas at risk, or to signal certain events, or again to transfer information between two places. In all these situations the possibility of transferring the desired information incorrectly is obviously a great constraint. At times, in fact, degradation of the information transmitted may depend on system malfunctioning and may have very serious consequences. In such cases it is possible to use the proposed techniques, for example forcing the transmitter with a dither signal with a suitable *PDF*. Obviously, the signal received will contain not only the associated information at the carrier signal but also spurious harmonics due to the noise signal. To filter these harmonics it will therefore be necessary to post-process the received signal.
As a second example, Figure 14 is a scheme of a distance measuring system using either an acoustic or an optical sensor. Using systems like the one illustrated in

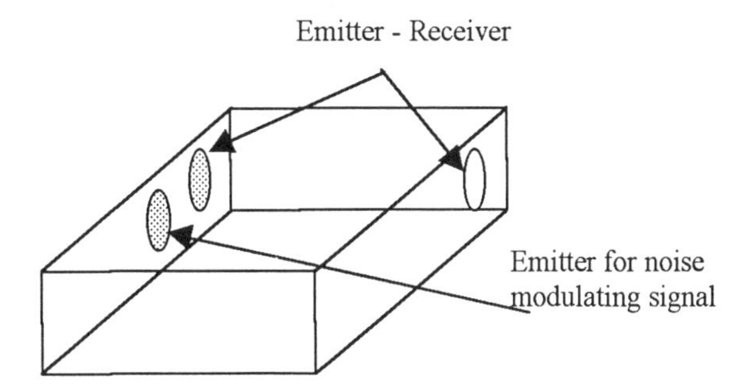

Figure 13. A pair of optical sensors for the remote transfer of information.

Figure 14, for example, a signal with known properties is normally transmitted and alterations in it are then associated with the desired distance measurement. The problems due to the threshold are reflected in the extension of the device's measurement range. Basically, when the distance to be measured is great, the receiver can no longer process the information contained in the transmitted signal. The characteristic of these sensors can be summarised as in Figure 15. It will be recalled that in the previous example the threshold error prevented the transfer of signals with a low amplitude. Although the problem here is different, it is possible to identify a common feature: the incapacity of some systems to process signals with a low amplitude. From the point of view of the receiver, it makes no difference whether these signals already have a low amplitude or obtain one due to absorption by the transmission medium they travel on. A way to extend the measurement range could be to eliminate the non-linearity by adding a noise signal and then processing the sensor's output signal. Here again, added-noise techniques can solve the problem.

In the two examples given, we have deliberately not dealt with the problem of what kind of noise should be used, what noise values will provide optimal working conditions, or how these optimal values can be determined. These topics will be dealt with in the following sections.

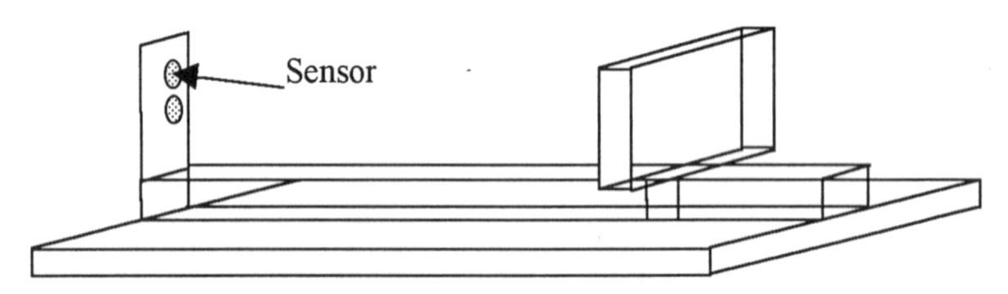

Figure 14. A system for distance measurement.

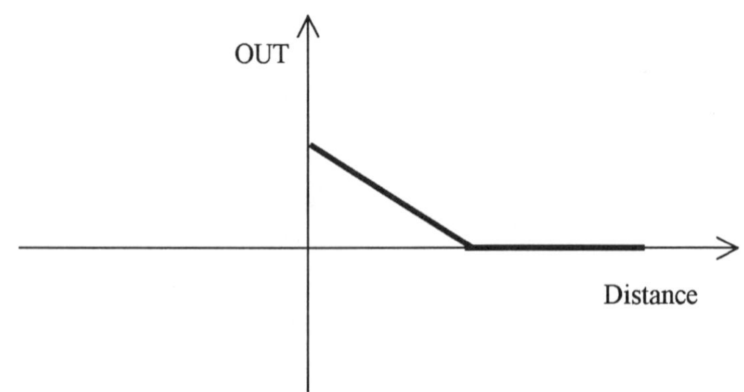

Figure 15. Characteristic of a distance measurement device.

3.4.3 THE NOISE OPTIMISATION PROCEDURE

The concept of stochastic modulation requires optimal noise signal parameters to be determined.

In section 3.2 we introduced a scheme of the procedure to determine the link between the optimal noise variance and the system parameters. When dealing with noise-added bistable systems this procedure led to the determination of iso-amplitude curves and the design of a system to control the optimal noise variance.

The deductive process which gave rise to this procedure is of general validity and can therefore also be applied to quasi-linear systems. The main differences can be summarised as follows

- unlike bistable systems, as noise-added quasi-linear devices are less affected by parametric variations, implementation of the control strategy is less complex;
- a complex output signal post-processing stage is required in quasi-linear systems. In bistable systems there is no post-processing stage because such systems naturally filter the high-frequency components. In quasi-linear systems, on the other hand, this section is necessary to assure that the undesirable noise components that the system transfers to the output will be filtered out.

To return to the procedure outlined in Figure 1, the first step is to determine the index to be optimised. We will then deal with the link between the optimal noise variance and the characteristics of the system.

In quasi-linear systems, as in bistable systems, to determine the index to be optimised according to the amplitude of the stochastic signal, a probabilistic approach was used. At times, this kind of approach has been used to study phenomena connected with dithering [7-9].

The possibility of a sub-threshold signal being able to force the system coincides with the probability that the amplitude of fluctuations in the stochastic signal will be contained in a range close to the threshold. This probability can be expressed as follows [15]

$$V=P(\lambda S < N(t) < \lambda S+\varepsilon) \qquad (3.27)$$

where

V represents the index to be optimised, taking into account the probability that fluctuations in the stochastic forcing signal will be able to force the system;

S is the system threshold;

$N(t)$ is the stochastic forcing signal;

ε defines the threshold area involved in fluctuations in the stochastic forcing signal;

$\lambda \in [0,1]$ represents the percentage of threshold reduction.

The latter parameter is of fundamental importance in practical use of these techniques because it makes it possible to specify a minimum threshold for the device to function, which does not necessarily have to correspond to the physical threshold of the device. So as not to increase excessively the noisiness of the system or the complexity of the post-processing stage, in fact, the decision is often made not to work in optimal operating conditions, i.e. with the maximum threshold error reduction, but in conditions which will meet the specifications required by the application. As with dithering theory [7-9], we determine the value of this probability using stochastic forcing signals with a *uniform* and *Gaussian* distribution.

Noise with a null-average *Gaussian* density and a variance σ^2 has the following probability density

$$f(x) = \frac{1}{\sigma} g(\frac{x}{\sigma}) = \frac{1}{\sigma\sqrt{2\pi}} e^{\frac{-x^2}{2\sigma^2}} \tag{3.28}$$

Likewise, a uniformly–distributed signal with an amplitude D has the following probability density

$$g(x) = \begin{cases} 1/D \text{ if } -D/2 < x < D/2 \\ D \quad \text{ if } x < -D/2; x > D/2 \end{cases} \tag{3.29}$$

as shown in Figure 16. For noise with a *Gaussian PDF*, expression (3.27) can be written as [16]

$$V_G = P(\lambda S < x < \lambda S + \varepsilon) = P(x < \lambda S + \varepsilon) - P(x < \lambda S) = G(\lambda S + \varepsilon/\sigma) - G(\lambda S/\sigma) \tag{3.30}$$

The evolution of this quantity with varying λ and σ values is shown in Figure 17.
In the case of noise with a *uniform* distribution, the probability V_u takes the following form

$$V_u = \begin{cases} \frac{\varepsilon}{D} & if \quad D/2 > \lambda S + \varepsilon \\ \frac{D/2 - \lambda S}{D} & if \quad \lambda S < D/2 < \lambda S + \varepsilon \\ 0 & if \quad D/2 < \lambda S \end{cases} \tag{3.31}$$

corresponding to schemes *a*, *b* and *c* in Figure 18. Figure 19 shows the evolution of the quantity V_u.

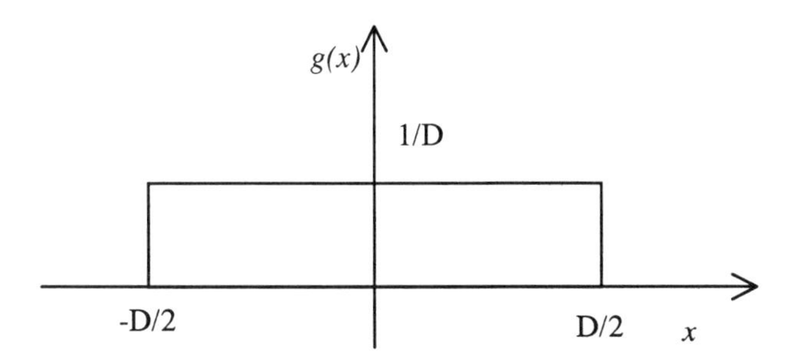

Figure 16. The Probability Density Function for uniformly-distributed noise.

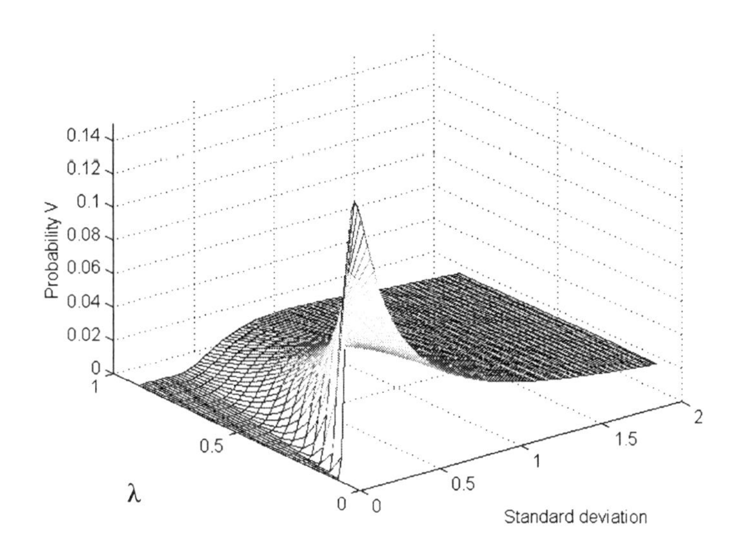

Figure 17. The evolution of the V_G quantity with varying λ and σ values.

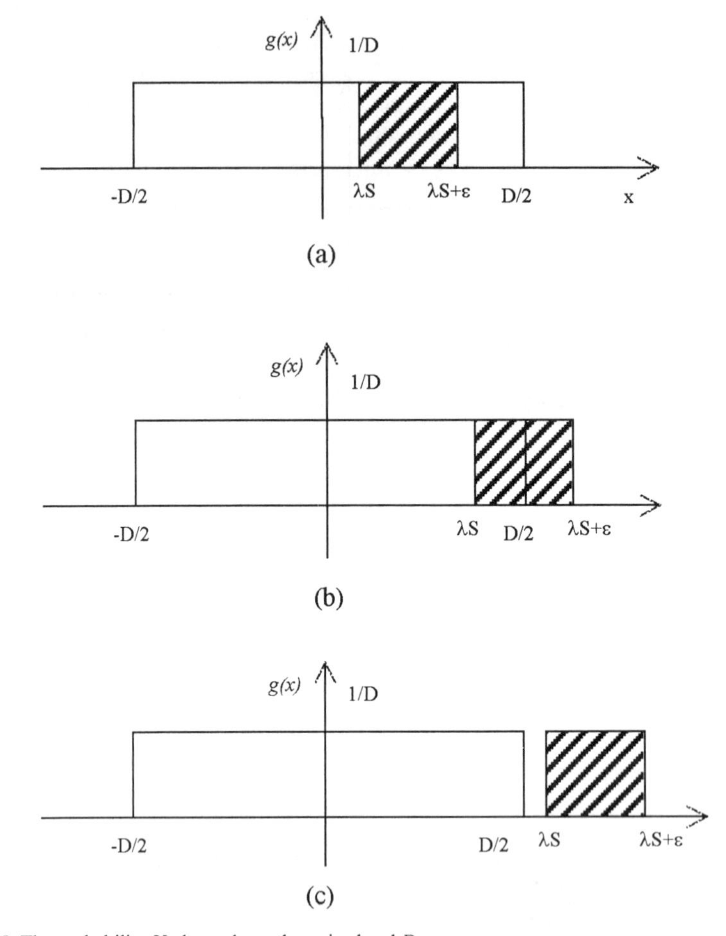

Figure 18. The probability V_u depends on the noise level D.

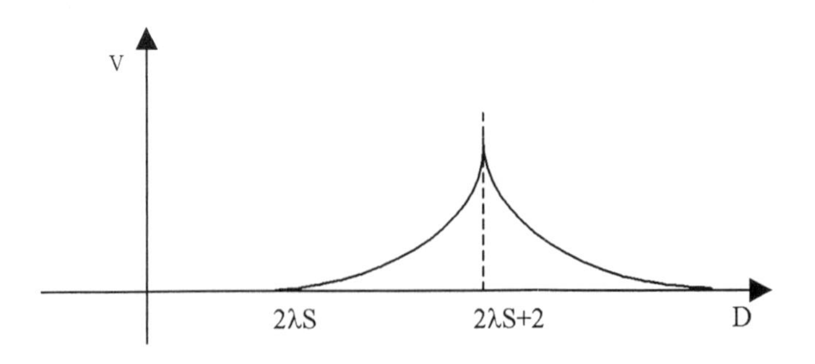

Figure 19. Evolution of the quantity V_u.

3.4.4 THE DETECTION OF AN OPTIMAL NOISE VARIANCE VALUE

The optimal noise variance, σ_{opt}, which will assure the desired system performance is the value which maximises the probability Vu or V_G, according to whether the noise has a *Gaussian* or *uniform* distribution.

With *Gaussian* noise the evolution of V_G for varying λ and σ values is as shown in Figure 17.

It is straightforward to demonstrate that this probability reaches a maximum when

$$\sigma^{G}_{opt} = \lambda S \tag{3.32}$$

From the evolution of V_u as a function of σ in Figure 19, it can be seen that the D_{opt} value with which V_u reaches a maximum is

$$D_{opt}=2\lambda S+2\varepsilon \tag{3.33}$$

As the following relation holds

$$\sigma^{u} = \frac{D}{\sqrt{12}} \tag{3.34}$$

we have

$$\sigma^{u}_{opt} = \frac{\lambda S + \varepsilon}{\sqrt{3}} \tag{3.35}$$

If $\lambda=1$ and ε is negligible, the results coincide with those given in the literature regarding dithering. This confirms the reliability of the deductions and results presented.

If the physical threshold S' is not directly comparable with the standard deviation σ of the stochastic signal, the quantity S in the probability V_α, $\alpha \in [G, u]$, has to be interpreted as a function of the system's physical threshold S'

$$S=f(S',p_1,p_2,...p_n) \tag{3.36}$$

where p_1, p_2, $...p_n$ represent generic magnitudes that may affect the system being considered.

Taking account of the function $f(\bullet)$, whose task is to transform S' to S, equations (3.32) and (3.35) will take the following form

$$\sigma^{G}_{opt} = \lambda S = f(S',p_1,p_2,...p_n) \tag{3.37}$$

$$\sigma^{u}_{opt} = \frac{\lambda f(S',p_1,p_2,\dots p_n)+\varepsilon}{\sqrt{3}} \quad (3.38)$$

Observe that when equations (3.37) and (3.38) are calculated with $f=I$ (identity function), they are the same as (3.32) and (3.35). With the optical sensor described in section 3.4.2 for example, the physical threshold of the system is represented by the maximum distance at which the sensor can detect the presence of the target, $S'=d_{max}$. If the target is placed at a greater distance than d_{max}, the receiver can no longer process the information contained in the return signal. For these devices there is a law associating the distance measured with the sensor's output voltage: $V_r=g(d)$. Having calculated the maximum measurable distance, $S'=d_{max}$, it is possible to obtain the corresponding lowest value of the return signal that can be detected: $V_{min}=g(d_{max})$.
V_{min} is the minimum voltage the receiver is able to process and therefore represents the threshold, S, of the system comprising the sensor, the transmission channel and the target.
It should, however, be pointed out that if the systems being studied have a threshold that cannot be expressed in terms of a quantity commensurable with the nature of the noise level, the above presented approach cannot be used. Infact, although the considerations about improving the devices operating conditions still hold, lack of knowledge of the function $f(\bullet)$ means that it is impossible to apply the technique to determine the optimal noise variance value, σ.
In such cases it will be necessary to use an index of an experimental nature to qualify the behaviour of the system as a function of the level of added noise.
Experimental examples of the application of the methods presented will be discussed in Part 2.

References

[1] L. Gammaitoni, F. Marchesoni, E. Menichella-Saetta, and S. Santucci, "Stochastic Resonance in Bistable Systems", *Phys. Rev. Lett. 62, 1989, 49.*

[2] Roberto Benzi, Alfonso Sutera and Angelo Vulpiani, "The mechanism of stochastic resonance", *J. Phys. A: Math. Gentile. 14, 1981, L453.*

[3] S. FAUVE and F. Heslot, "Stochastic Resonance in a Bistable Systems", *Phys. Lett. 97A,1983, 5.*

[4] F. Marchesoni, E. Menichella-Saetta, M. Pochini and S. Santucci, "Analog simulation of underdamped system driven by colored noise: Spectral densities", *Phys. Rev. A 37, 1988, 3058.*

[5] Francois Chapeau-Blondeau, Xavier Godivier, and Nicolas Chambet, "Stochastic resonance in a neuron model that transmit spike trains", *Phys. Rev. E 53, 1996, 1273.*

[6] L. Gammaitoni, M. Martinelli, and L. Pardi, "Observation of Stochastic Resonance in Bistable Electron-Paramagnetic-Resonance Systems", *Phys. Rev. Letters, Vol.67, N. 13, 1991, 1799.*

[7] Paolo Carbone and Dario Petri, "Effect of Additive Dither on the Resolution of Ideal Quantizer", *IEEE Transaction on instrumentation and measurement Vol. 43, N. 3, 1994, 389.*

[8] Paolo Carbone, Claudio Narduzzi, and Dario Petri, "Performance of Stochastic Quantizer Employing Nonlinear Processing", *IEEE Transaction on instrumentation and measurement Vol. 45, N. 2, 1996, 435.*

[9] Luca Gammaitoni, "Stochastic Resonance and the dithering effect in threshold physical systems", *Phys. Rev. E 52, 1995, 4691.*

[10] L. Gammaitoni, P. Hanggi, P. Jung, F. Marchesoni, "Stochastic Resonance", Rev. Of Modern Physics vol. 70, 1, 1998.

[11] B.Andò, S. Baglio, S. Graziani, N. Pitrone., 1999, "Optimal improvement in bistable measurement device perfromance via stochastic resonance." INT. J. ELECTRONICS, vol 86, n. 7.

[12] B.Andò, S. Baglio, S. Graziani, N. Pitrone, "A probabilistic approach to the threshold error reduction theory in bistable measurement devices", *IMTC98, S. Paul, Minnesota, 1998.*

[13] B. Andò, S. Baglio, S. Graziani, N. Pitrone, "Characterisation of threshold error via stochastic resonance", in: IMEKO '97, Helsinki, 1997

[14] B. Andò, S. Baglio, S. Graziani, N. Pitrone, "Virtual instruments with low threshold error based on stochastic resonance theory", *SICICA '97, Annecy, France, 1997.*

[15] B.Andò, S. Baglio, S. Graziani, N. Pitrone, " Threshold error reduction in linear measurement devices by adding noise signal", *IMTC98, S. Paul, Minnesota, 1998.*

[16] A. Papoulis, "*Probability, Random Variables, and Stochastic Process*", McGRAW-HILL BOOK COMPANY.

[17] B. Andò, S. Baglio, S. Graziani, N. Pitrone, "A system for the implementation of noise added System driving", IMTC'99, Venezia, 1999.

[18] Ernest O. Doebelin, "*Measurement Systems*", *McGRAW-HILL BOOK COMPANY*, third edition, 1985.

4 ANALOG NOISE GENERATION VIA NON-LINEAR DEVICE

4.1 Introduction: the problem of noise generation

As mentioned previously, some systems work better in the presence of noise and a number of natural phenomena can be explained if the presence of noise is taken into account. This surprising characteristic of noise-added systems can be exploited to improve the performance of several systems, ranging from electronic circuits to a generic measuring device.
The problem of noise tuning so as to achieve optimal system performance has been dealt with in the previous chapters. Of course, an *on-line* application of noise-added systems requires the use of a suitable noise generator. In this chapter the possibility of using a Cellular Neural Network as a noise generator is investigated. It is well known that *CNN*s can have very complex dynamics and are analog devices that are capable of working on line as signal generators. In the following sections a *CNN* implementing the *Chua* system generating both *Gaussian* and *uniform* white noise will be discussed, and suitable techniques for *CNN* parameter estimation will be investigated.
In the field of electrical and electronic engineering various types of systems capable of generating signals with well-defined characteristics have been proposed and analysed. Recently, for instance, research has been carried out into non-linear generators [1] to replace complicated generators of pseudo-random numerical sequences [2], cryptosystems [3] and *1/f* noise generators [4].
It should also be pointed out that it is possible to control the dynamic evolution of the state of these non-linear systems by acting on the value of their structural parameters [5]. It would therefore appear possible to construct analog networks for the generation of signals with certain statistical and spectral characteristics.

4.2 Non-linear systems for the analog generation of noise signals

A stochastic signal is disordered and aleatory, unpredictable and non-deterministic; its evolution cannot be established a priori and never repeats itself. However, as reported in Chapter 1, it does possess certain features that allow us to classify it.
If, for example, we analyse the noise introduced in any communication system by the transmission channel, or the noise recorded in any analog network or, more

simply, the noise at the ends of a conductor coil at a temperature other than absolute zero, we observe that this particular type of noise is distributed around a mean value, with a *Gaussian* distribution probability: *Gaussian* noise is one of the most common examples.
A stochastic signal can also be characterised by mean of its spectral power density, a function that shows how the total power of the signal is distributed between components of different frequencies. More specifically, if we analyse the spectral power density of what is called white *Gaussian* noise, we observe that its frequency remains constant: this is accounted for by the fact that all the components at different frequencies are present in equal measure. Conversely, by filtering white *Gaussian* noise we obtain non-white noise, the spectral power density of which is no longer constant but varies according to the type of filtering applied to the original signal.
On the basis of what we have said so far, we will now assess the possibility of setting suitable sets of parameters for a non-linear system, the *Chua* system, in such a way that its statistical and spectral characteristics will approximate those we are looking for. More specifically, we will search for parametric configurations that will give the system output a *Gaussian* statistical distribution and a *uniform* one. A *Chua* circuit is the best-known non-linear system for the generation of signals with complex dynamics [6,7]. By varying its parametric configuration, in fact, it is possible to reproduce dynamics with completely different characteristics, ranging from periodic to chaotic behaviour. In this context it would seem plausible to search for a set of parameters that will provide the system with a dynamic similar to a stochastic signal of known properties. In addition, as mentioned previously, the possibility of analog implementation makes such a system highly interesting from a practical point of view.

4.2.1 AN OVERVIEW OF THE CHUA CIRCUIT TOPOLOGY

A general scheme of the *Chua* circuit is shown in Figure 1. The network is made up of four linear elements (two capacitors, an inductor and a linear resistor) and a non-linear resistor [7]. Worth mentioning is the fact that the resistor R is inserted in such a way as to not create a parallel between the two capacitors, thus increasing the order of the system from two to three [8]. The functioning of the network is described by the following system of differential equations

$$
\begin{aligned}
C_1 \frac{dv_1}{dt} &= \frac{1}{R}(v_2 - v_1) - f(v_1) \\
C_2 \frac{dv_2}{dt} &= -\frac{1}{R}(v_2 - v_1) + i_3 \\
L \frac{di_3}{dt} &= -v_2 + R_3 i_3
\end{aligned}
\tag{4.1}
$$

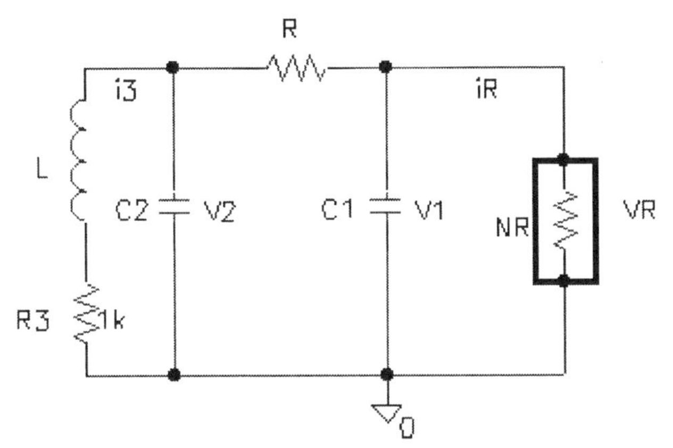

Figure 1. The Unfolded *Chua* circuit.

where the quantity

$$f(v_R)=G_b v_R+0.5(G_a-G_b)\left\{\left|v_R+B_p\right|-\left|v_R-B_p\right|\right\} \tag{4.2}$$

identifies the three segments of the voltage-current feature of the non-linear resistor with slopes G_a and G_b, and slope change points placed at $v_R = -B_p$ and $v_R = B_p$ respectively. The voltage-current feature of the non-linear resistor, implemented by a *Chua* diode, is shown in Figure 2. Note that the system being analysed is autonomous in the sense that there are no forced external signals in the system.
Changing the variables, the state equations (4.1) can be transformed into the following a-dimensional form [7]

$$\begin{aligned}\frac{dx}{dt}&=k\cdot\alpha\cdot(y-f(x))\\ \frac{dy}{dt}&=k\cdot(x-y+z)\\ \frac{dz}{dt}&=-k\cdot\beta\cdot y-\gamma\cdot k\cdot z\end{aligned} \tag{4.3}$$

where

$$f(x)=m_1 x+0.5\cdot(m_0-m_1)\cdot(\left|x+1\right|-\left|x-1\right|) \tag{4.4}$$

and

$$x \equiv \frac{v_1}{B_p},\ y \equiv \frac{v_2}{B_p},\ z \equiv i_3 \cdot \left(\frac{R}{B_p}\right)$$

$$\alpha \equiv \frac{C_2}{C_1},\ \beta \equiv \frac{R^2 C_2}{L},\ k \equiv 1/(RC_2) \tag{4.5}$$

$$m_0 \equiv RG_a + 1,\ m_1 \equiv RG_b + 1,\ \gamma = \frac{C_2 R_3 R}{L}$$

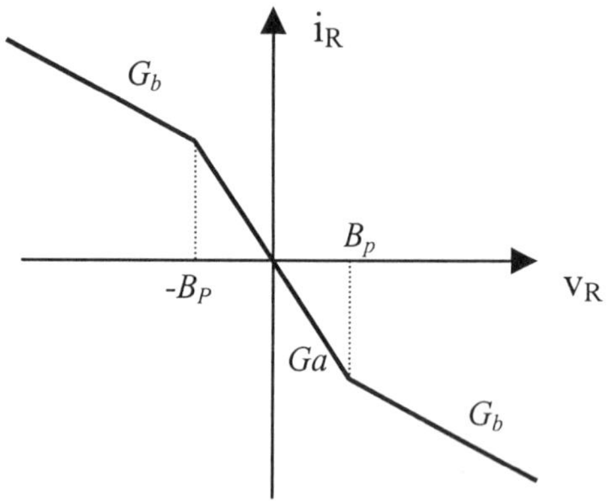

Figure 2. Voltage-current characteristic of the non linear resistor in *Chua* circuit.

A "zoo" of attractors generated by these equations and obtained by varying the main parameters α, β, γ, m_0, m_1 is to be found in [7-8].
To observe the behaviour of the *Chua* system it is possible to use numerical simulations environments or analog implementations.
In the following sections both techniques will be used. Simulators will be used to identify the most suitable set of parameters to give the system the required dynamics, followed by analog implementation of the configurations considered to be the most interesting.

4.2.2 NUMERICAL SIMULATION OF THE CHUA CIRCUIT

The first step is to determine the parameters of the *Chua* circuit identifying that set of chaotic attractors that have a *uniform* or *Gaussian PDF*. To do so it is necessary to use a simulation method that will emulate the behaviour of the *Chua* cell, or rather of the system of differential equations (4.1), as its parameters vary. It is, for example, possible to simulate the system of differential equations by using the *Matlab®* *SIMULINK* simulation environment.

A block diagram of the system is shown in Figure 3.
In order to speed up the simulation process when the system parameters α and β vary, it is possible to write a *Matlab®* script. This routine cyclically varies at least two of the system parameters (4.2) and orders the signals *x(t)* and *y(t)* in matrix form.

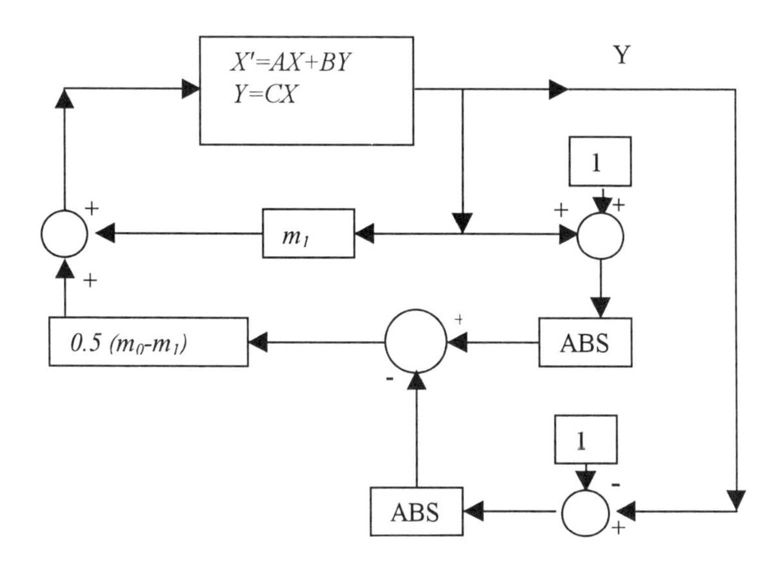

Figure 3. *SIMULINK®* representation of the *Chua system* (4.3).

The values assigned to the remaining parameters are as follows

$k=1$
$m_0=-0.1428571;$
$m_1=0.2857142;$
$x(0)=y(0)=z(0)=0.1.$

The output variables analysed, *x(t)* and *y(t)*, correspond to the voltages V_{c1} and V_{c2} at the terminals of the two capacitors C_1 and C_2 in the *Chua* circuit shown in Figure 3.
The next step is to analyse the signals obtained so as to evaluate their statistical and spectral characteristics.
It is therefore necessary to introduce statistical methods that will allow us to calculate how close a time sequence is to a *uniform* or a *Gaussian* amplitude distribution.

4.2.3 A GAUSSIAN LIKE NOISE GENERATION

In order to determine which of the signals obtained with varying α and β values is closest, in probabilistic terms, to a *Gaussian* noise, it is possible to use the χ^2 method [9]. Details of implementation of this method are given in the Appendix B and can be found in a number of texts on statistical analysis.

Evaluating the index χ^2 for all the sequences generated by simulation, it is possible to find the one for which χ^2 takes the lowest value and thus the one which is likely to be the closest to a *Gaussian* noise.

Some results of an investigation of this kind are given in Figure 4, which shows how the index χ^2 evolves as the parameters α and β, vary. In this case, patterns of 500 data items divided into 22 groups were analysed. From this we can determine the α and β values with which the index χ^2 is minimum ($\chi^2=10.4$). Analysis of the test evaluation tables given in the Appendix B shows that with the parameter values selected the network generates a signal, V_G, whose distribution is *Gaussian*-like with a confidence of 95%.

The parameter values optimising the functioning of the system are the following

$\alpha=9.6;$
$\beta=15.2;$
$k=1;$
$m_0=-0.1428571;\ m_1=0.2857142;$

the initial conditions being

$x(0)=y(0)=z(0)=0.1$

Evolution of the signal V_G, corresponding to the voltage at the terminals of capacitor C_2 in the network shown in Figure 1, is shown in Figure 5. Figures 6 and 7 give the *PDF* and spectrum of the voltage V_G. Note that the noise signal obtained has satisfactory characteristics.

4.2.4 A UNIFORM LIKE NOISE GENERATION

In generation of a signal with a *uniform PDF*, the search for the optimal parameter configuration again has to be made using the data sequences generated by a simulation. As before, it is necessary to use an analytical method that will find the time variables $x(t)$ and $y(t)$ whose amplitude distribution is *uniform*. Again, we can use the χ^2 test.

A typical trend for the index χ^2, with varying α and β values, is shown in Figure 8. In this case, performing the test on 500 data items divided into 20 groups the minimum χ^2 value found was 6.5. This means that in the family of vectors $x(t)$ and $y(t)$ there exists a vector with a *PDF* that can be compared to that of *uniform* noise with a 95% probability.

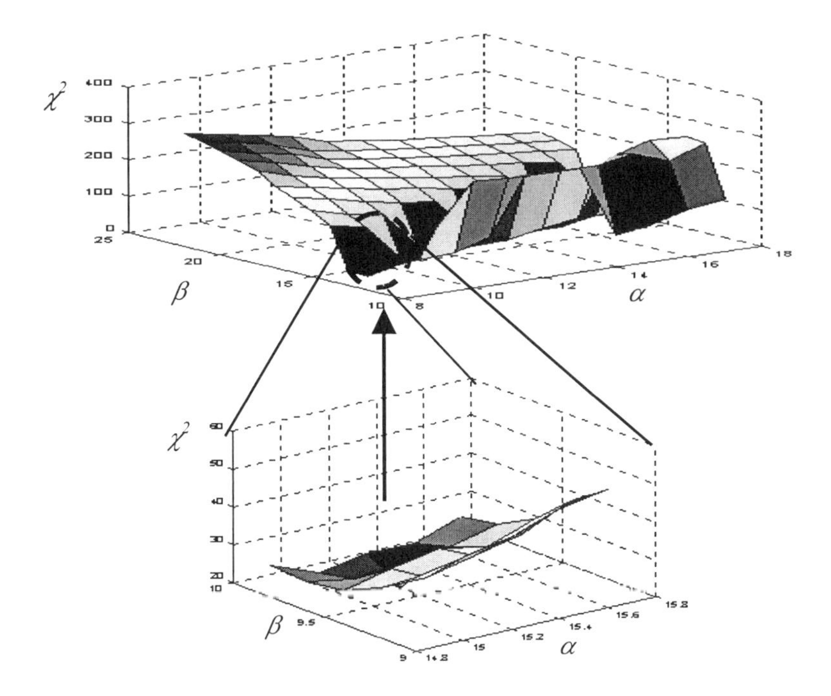

Figure 4. The index χ^2 for all the sequences generated by simulation in case of a white *Gaussian* like signal investigation. The minimum of the χ^2 quantity is obtained when $\alpha = 9.6$ and $\beta = 15.2$.

The parameter values optimising functioning of the system in this case are

$\alpha=9;$
$\beta=14.2;$
$k=1;$
$m_0=-0.1428571;\ m_1=0.2857142;$

the initial conditions being

$x(0)=y(0)=z(0)=0.1$

Evolution of the sequence selected is shown in Figure 9. Here again, it corresponds to the voltage V_c at the ends of the capacitor C_2. Figures 10 and 11 give the *PDF* and spectrum of the signal.

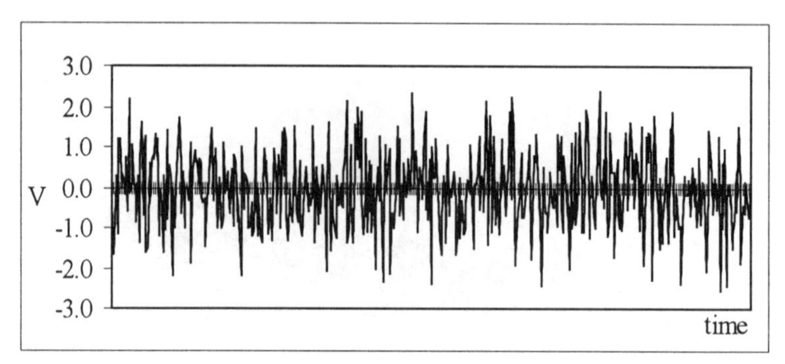

Figure 5. Evolution of the signal V_G, obtained by *Simulink*® simulation, in case of white *Gaussian* like signal investigation.

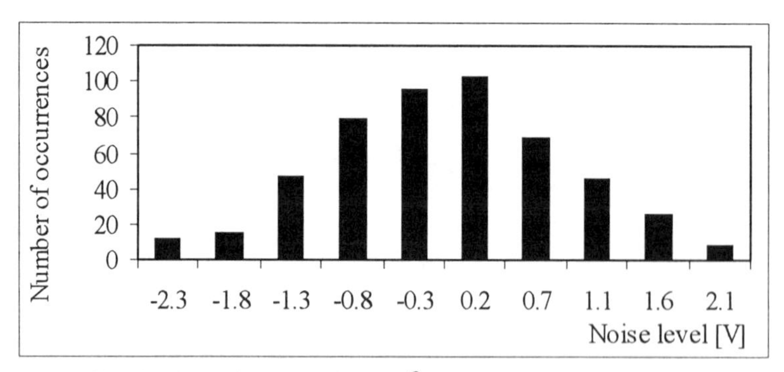

Figure 6. *PDF* of the signal V_G, obtained by *Simulink*® simulation, in case of white *Gaussian* like signal investigation.

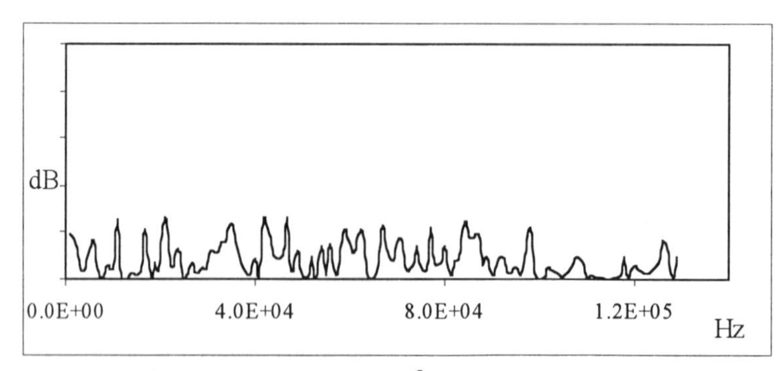

Figure 7. *FFT* of the signal V_G, obtained by *Simulink*® simulation, in case of white *Gaussian* like signal investigation.

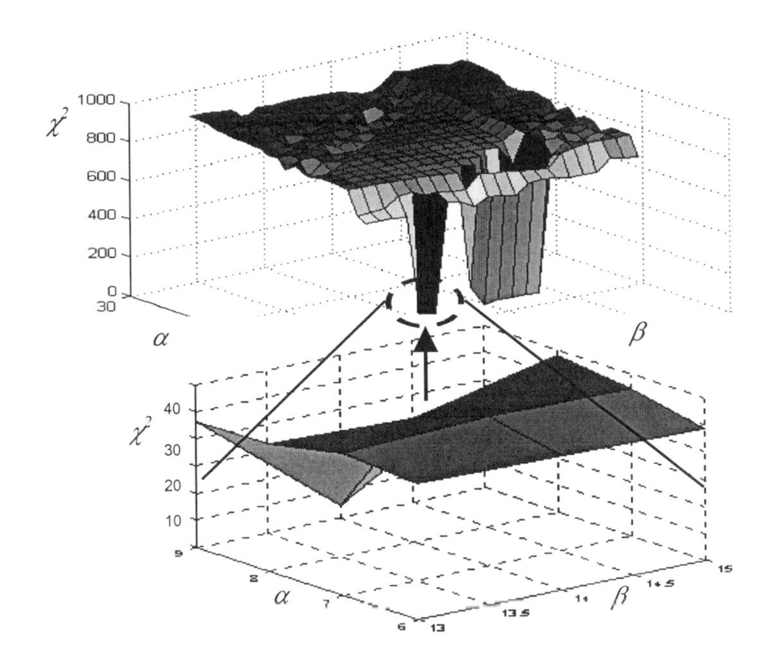

Figure 8. The index χ^2 for all the sequences generated by simulation in case of white *uniform* like signal investigation. The minimum of the χ^2 quantity is obtained when $\alpha = 9$ e $\beta = 14.2$.

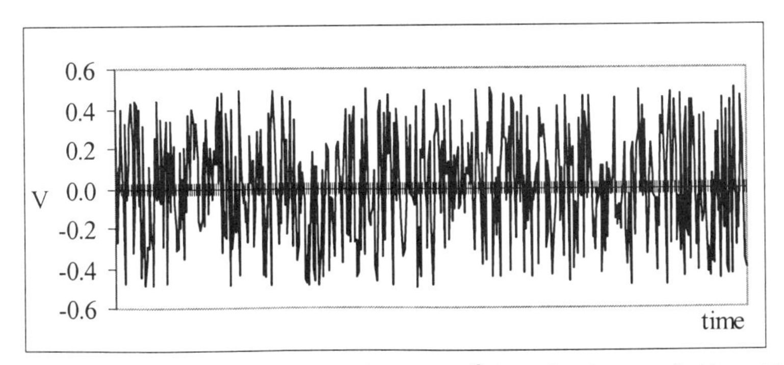

Figure 9. Evolution of the signal V_U, cbtained by *Simulink*® simulation, in case of white *uniform* like signal investigation.

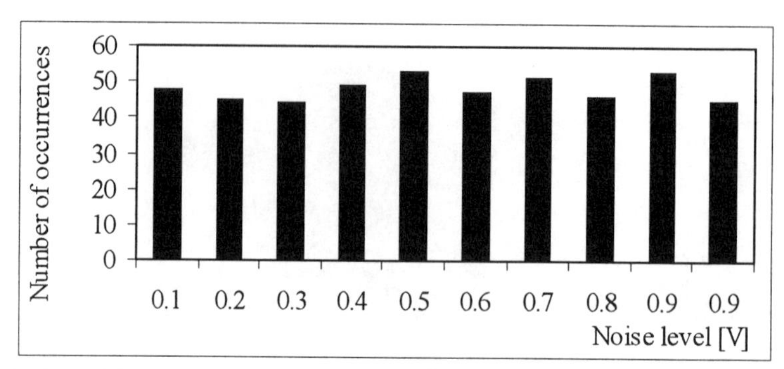

Figure 10. *PDF* of the signal V_U, obtained by *Simulink®* simulation, in case of white *uniform* like signal investigation.

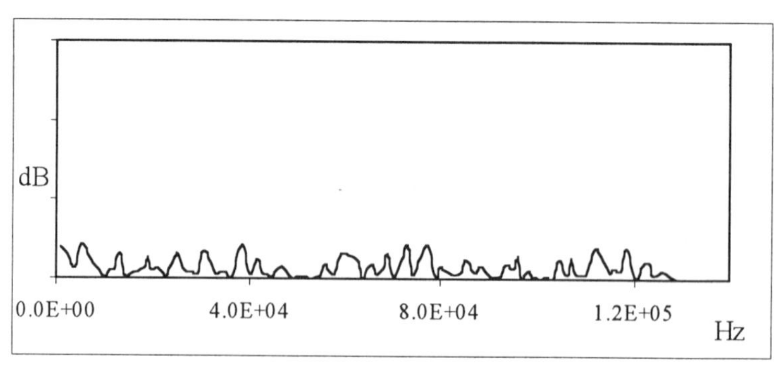

Figure 11. *FFT* of the signal V_U, obtained by *Simulink®* simulation, in case of white *uniform* like signal investigation.

4.3 An Overview of Cellular Neural Networks (CNNs)

The use of *CNN*s to generate chaotic signals was first introduced in 1988 by Chua and Yang [10]. A *CNN* is typically made up of locally connected elementary cells. If first order cells are considered, they are described by equations of the following kind

$$\begin{aligned} \frac{dx_j}{dt} &= -x_j + a_j y_j + G_o + G_s + i_j \\ y_j &= 0.5 * (|x_j + 1| - |x_j - 1|) \end{aligned} \qquad (4.6)$$

where

j is the index of a generic cell, and x_j its state variable;
y_j is the output of the cell;
a_j is a constant parameter and i_j is a threshold value.

In equation (4.6) the terms G_o and G_s are the linear combination of the outputs and state variables of the neighbouring cells. The dynamic model of three fully connected generalised *CNN* cells is

$$\begin{aligned} \dot{x}_1 &= -x_1 + a_1 y_1 + a_{12} y_2 + a_{13} y_3 + \sum_{k=1}^{3} s_{1k} x_k + i_1 \\ \dot{x}_2 &= -x_2 + a_{21} y_1 + a_2 y_2 + a_{23} y_3 + \sum_{k=1}^{3} s_{2k} x_k + i_2 \\ \dot{x}_3 &= -x_3 + a_{31} y_1 + a_{32} y_2 + a_3 y_3 + \sum_{k=1}^{3} s_{3k} x_k + i_3 \end{aligned} \tag{4.7}$$

where s_{nm} are generic coefficients to be determined according to the complex model being described.

A simplified dynamic model of a *CNN* comprising three cells, which looks similar to the adimensional *Chua* system (4.3), is given in [11]

$$\begin{aligned} \frac{dx_1}{dt} &= -x_1 + a_1 y_1 + s_{11} x_1 + s_{12} x_2 \\ \frac{dx_2}{dt} &= -x_2 + s_{21} x_1 + s_{23} x_3 \\ \frac{dx_3}{dt} &= -x_3 + s_{32} x_2 + s_{33} x_3 \end{aligned} \tag{4.8}$$

By referring to the a-dimensional differential system (4.8) it is possible to create an analog model for each cell. Each cell is essentially made up of three blocks, as shown in Figure 12 [11].
Section B1 constitutes the non-linearity of the network; it uses the natural saturation of the output of the amplifier U_{2A}, with an appropriate choice of R5 and R6, when $|x_j|>1$. Resistors R9 and R10 are selected in such a way as to scale the output voltage $-y_j$ in the range [-1,1]. The conditions to be imposed are therefore

$$\begin{aligned} &R5/R6=V_{satA}/V_{satx} \\ &R6/R5=R10/(R9+R10) \end{aligned} \tag{4.9}$$

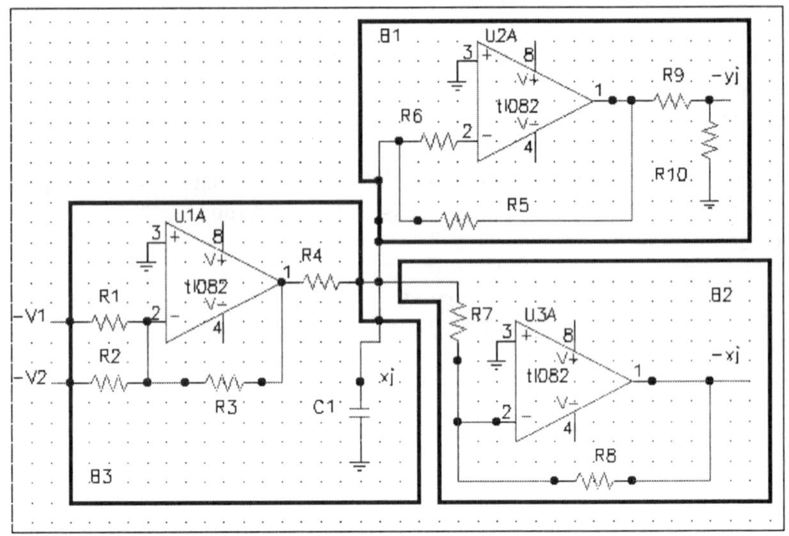

Figure 12. A circuit representation of a single cell.

where V_{satA} is the saturation voltage at the amplifier output and V_{satx} is the corresponding input voltage.
Section B2 is an inverting amplifier with a unit gain. We therefore have

$$R7=R8 \tag{4.10}$$

Finally, section B3 represents the heart of the cell. If the parallel between the input impedance of B1 and B2 is high and comparable with the block B3 output impedance, then blocks B1 and B2 do not load the capacitor C_l. This is the case if the following holds

$$(R7R6)\,/\,(R7+R6) >> R4 \tag{4.11}$$

In this case the cell's generic state equation is

$$C_J \dot{x}_j = -\frac{x_j}{R4} + \frac{R3}{R1R4} V_1 + \frac{R3}{R2R4} V_2 \tag{4.12}$$

This demonstrates that each of the equations in model (4.8) can be implemented using the circuit scheme shown in Figure 12. Equations 4.12 are equivalent to (4.8) for suitable value of the circuit parameters and with the following condition

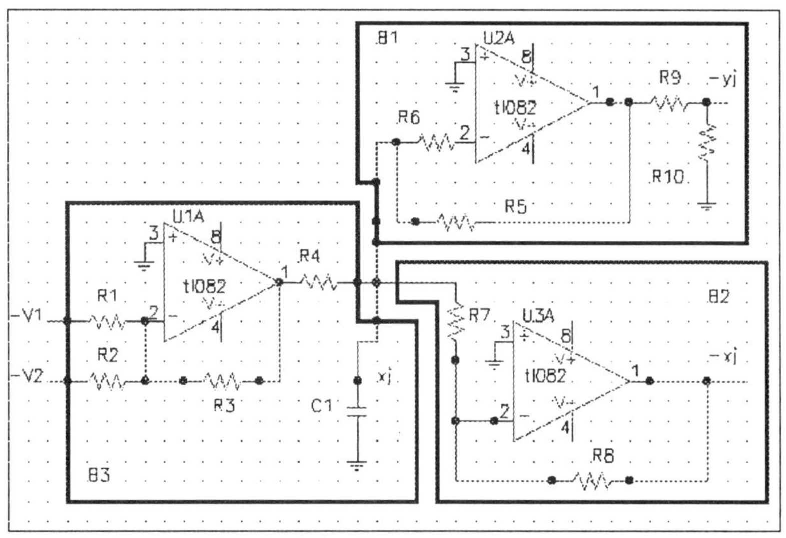

Figure 12. A circuit representation of a single cell.

where V_{satA} is the saturation voltage at the amplifier output and V_{satx} is the corresponding input voltage.
Section B2 is an inverting amplifier with a unit gain. We therefore have

$$R7=R8 \tag{4.10}$$

Finally, section B3 represents the heart of the cell. If the parallel between the input impedance of B1 and B2 is high and comparable with the block B3 output impedance, then blocks B1 and B2 do not load the capacitor C_1. This is the case if the following holds

$$(R7R6)/(R7+R6) >> R4 \tag{4.11}$$

In this case the cell's generic state equation is

$$C_J x_j = -\frac{x_j}{R4} + \frac{R3}{R1R4}V_1 + \frac{R3}{R2R4}V_2 \tag{4.12}$$

This demonstrates that each of the equations in model (4.8) can be implemented using the circuit scheme shown in Figure 12. Equations 4.12 are equivalent to (4.8) for suitable value of the circuit parameters and with the following condition

$V_1=y_1,\ V_2=x_2$ (4.13)
for the first cell;

$V_1=x_1,\ V_2=x_3$ (4.14)
for the second cell;

$V_1=-x2,\ V_2=x_3$ (4.15)
for the third cell,

Occurrences of this conditions is schematised in Figure 13.

4.3.1 THE CHUA CIRCUIT IMPLEMENTED BY CNN

If we observe the differential system referring to the *Chua* network (4.3), we find analogies with system (4.7).
More specifically, if we set

$$a_1=\alpha(m_1-m_0); \quad s_{33}=1-\gamma; \quad s_{21}=s_{23}=1;$$
$$s_{11}=1-\alpha\, m_1; \quad s_{12}=\alpha; \quad s_{32}=-\beta; \tag{4.16}$$

we obtain the system (4.3) with x_1, x_2 and x_3 respectively equal to x, y and z in the system being examined.
Hence, the equations (4.12), referring to the circuit being considered, are equivalent to system (4.8), which in turn is equivalent to the a-dimensional differential *Chua* system (4.3), with the conditions given by (4.16).
As shown so far, therefore, the use of a *CNN* simplifies implementation of a *Chua* system, without the need to use inductors and non-linear resistors. Once a simple solution has been found for the analog implementation of noise generators, it is sufficient to determine the parameters of the cells $a_1, s_{11}, s_{12}, s_{21}, s_{23}, s_{32}, s_{33}$, using equation (4.12) on the values of the parameters α, β, γ, m_0, m_1, τ with which the system output is respectively *Gaussian* and *uniform*.

4.3.2 THE GAUSSIAN GENERATOR IMPLEMENTED BY CNN

It should be recalled that, as described previously, the parameter configuration of the *Chua* system for the generation of a signal with a *Gaussian PDF* is as follows

$\alpha=9.6; \beta=15.2;\ k=1;$
$m_0=-0.1428571;\ m_1=0.2857142;$

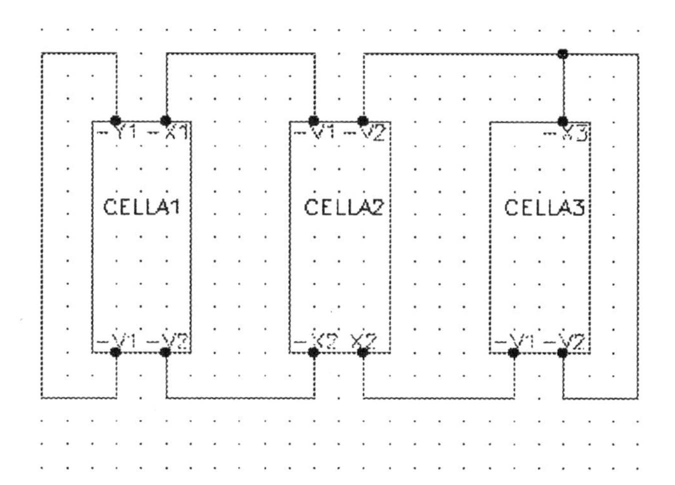

Figure 13. The *CNN* structure implementing the *Chua* system.

Referring to relations (4.16), in the *CNN* these parameters become:

$a_1=\alpha(m_1-m_0)=3.857;\ s_{33}=1;\ s_{21}=s_{23}=1;$
$s_{11}=1-\alpha\, m_1=-1.5714;\ \ s_{12}=\alpha=9;\ \ s_{32}=-\beta=-14.3;$

The *Chua* circuit for the generation of *Gaussian* noise was simulated in a *PSspice®* environment and implemented in hardware. Figures 14 and 15 show the evolution and distribution of the voltage V_{C2} generated via simulation and in the hardware implementation. The spectrum of the same signals is shown in Figure 16.
Comparison of the voltages produced experimentally and by simulation confirms the possibility of simple analog implementations for the generation of stochastic dynamics with known characteristics.

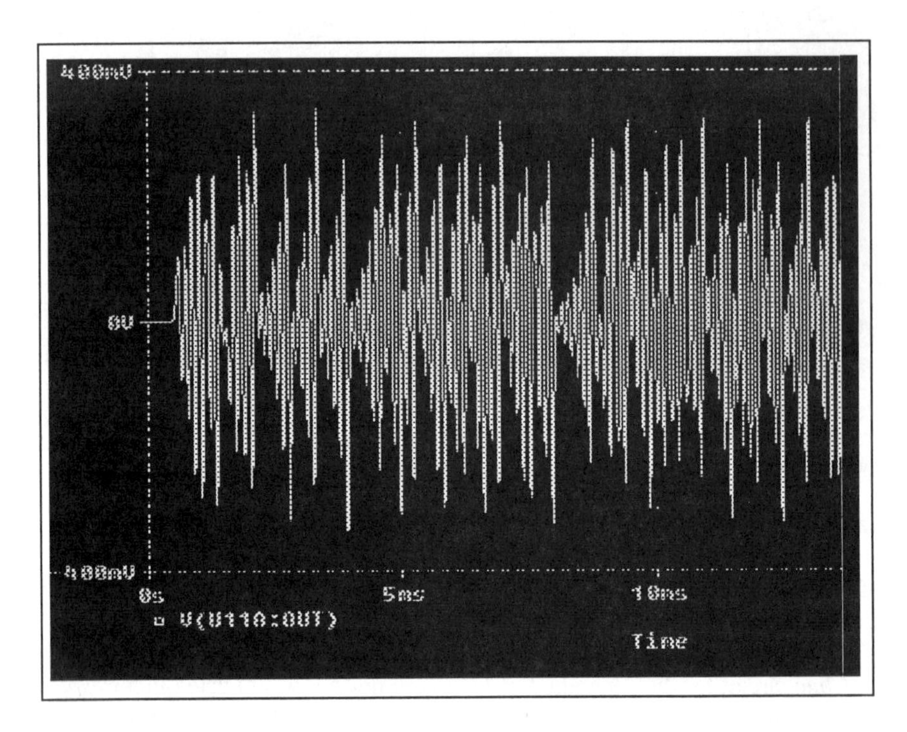

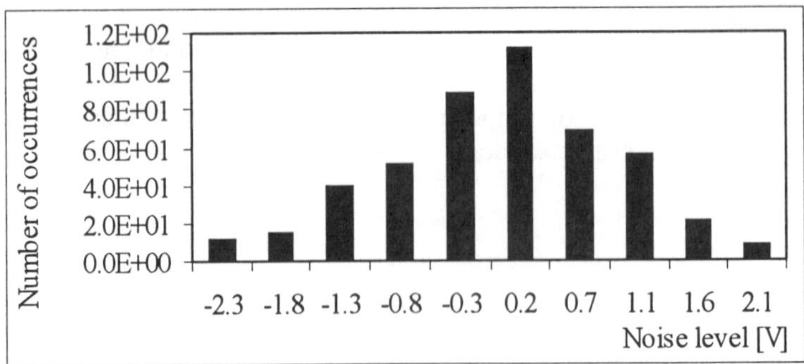

Figure 14. Evolution and *PDF* of the signal V_G, obtained by *PSspice®* simulation, in case of white *Gaussian* like signal generation via *CNN*.

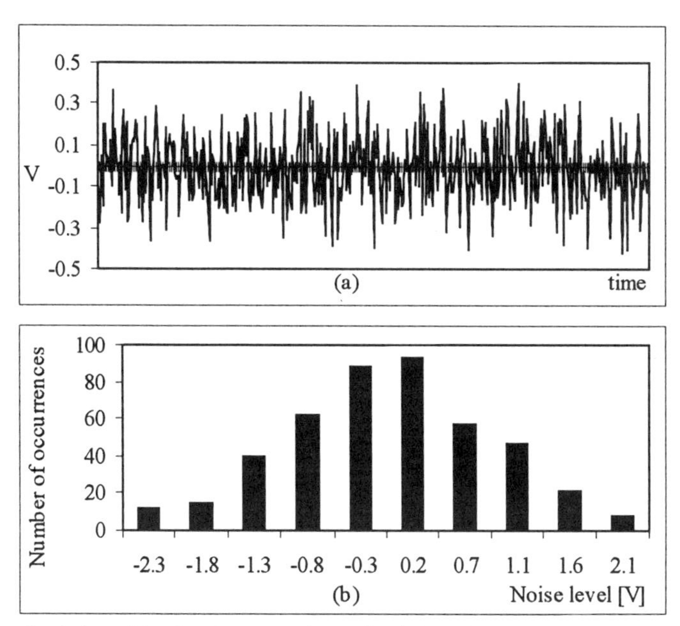

Figure 15. (a) Evolution of the signal V_G, experimentally obtained, in case of white *Gaussian* like signal generation via *CNN* and (b) the corresponding *PDF*.

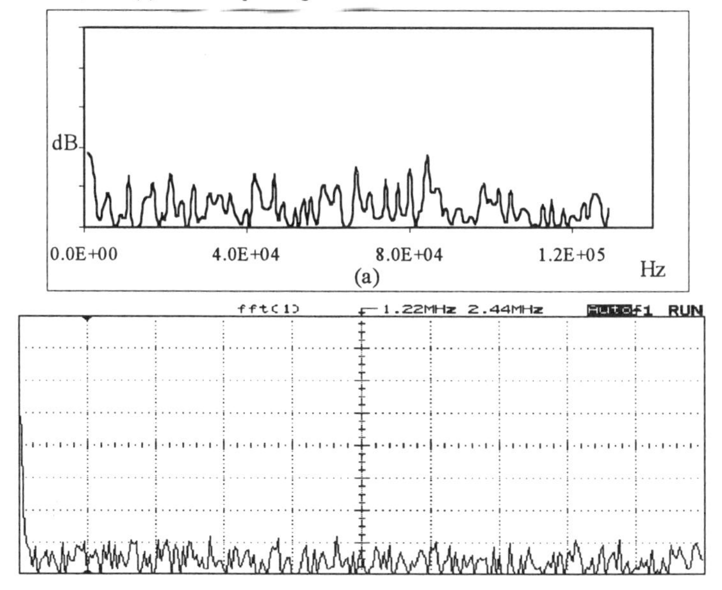

Figure 16. *FFT* of the signal *VG*, obtained by both (a) *PSspice®* simulation and (b) experimental set-up, in case of white *Gaussian* like signal generation via *CNN*.

References

[1] T. Kohda, K: Aihara, "Chaos in Discrete Systems and Diagnosis of Experimental Chaos", Trans. IEICE. Vol.E73, No.6 June 1990.

[2] Oishi, H. Inoue, *"Pseudo-Randum Number Generators and Chaos"*, Trans. IEICE. Vol.E65, No.9 September 1982.

[3] Habutsu, Y. Nishio, I. Sasase, S. Mori, "A Secret Key Cryptosystem Using a Chaotic Map", Trans. IEICE vol.E73, No.7 July 1990.

[4] Kohda, "An Electronic Noise Generator with 1/f Spectrum", IEEE Proc IS-CAS'85 pp.859-862, 1985.

[5] Inaba, T. Saito, S. Mori, "Chaotic Phenomena in a Circuit with a Negative Resistance and an Ideal Switch of Diodes", Trans. IEICE, vol.E70, No.8, pp.744-754, 1987.

[6] Madan, "Chua's Circuit: A Paradigm for Chaos", Singapore: World Scientific, 1993.

[7] O. Chua, "Global unfolding of Chua's circuit", IEICE Trans. Fundamentals, vol.E76-A, pp.704-734, May 1993.

[8] J.M.T. Thompson and H.B. Stewart, *"Nonlinear dynamics and chaos"*, John Wiley and Sons, 1989.

[9] Papoulis, *"Probability, Random Variables, and Stochastic Process"*, McGRAW-HILL BOOK COMPANY.

[10] L. O. Chua, L. Yang, "Cellular Neural Networks: Theory", IEEE Trans. Circuits Syst., vol.35, pp.1257-1272,1988.

[11] P.Arena, S.Baglio, L.Fortuna, G.Manganaro, "Chua's Circuit Can Be Generated by CNN Cells" IEEE Trans. Circuits Syst., vol.42, pp.123-125, February 1995.

PART 2 APPLICATIONS

5 APPLICATIONS

5.1 Introduction

In the last chapters it was emphasised that co-operation between a periodic signal and a suitable noise signal leads to dramatic changes in system performance. Both bistable and linear systems have been investigated and many studies have been carried out in order to assess the effects of noise addition.
When the *SR* condition is achieved, an input signal with an amplitude lower than the threshold value is able to force the system. This phenomenon can be interpreted as a *reduction in the system threshold level*, without any change in the input-output characteristic of the system. Such a technique is suitable when the information is transferred by using an input signal whose amplitude is unable to force the bistable system. *SR* theory allows for the detection of optimal noise parameter values [1-8].
On the other hand, linearity is the main property of a number of measuring systems though they show a certain non-linear behaviour when very low amplitude input signals are considered [9-13].
In the following sections, some examples of threshold reduction and system linearisation will be briefly introduced.

5.2 Threshold reduction in an electronic comparator

Noise-added theories can be adopted to develop a methodology which allows us to reduce the threshold error in measuring systems [4-8]. In this section the stochastic resonance phenomenon is applied to an experimental electronic comparator in order to select the system parameter values that will minimise the device threshold error.
In addition, the possibility of controlling system parameter fluctuations by means of an external "autotuning" control system is explored in order to enhance some metrological characteristics of the system.
The "autotuning" control system has been created using the *LabVIEW®* tool, produced by National Instrument. A virtual instrument is then proposed which combines the original analog instrument and the "autotuning" control system.
An important parameter for characterising the *SR* phenomenon, in bistable systems, is the Signal-to-Noise Ratio *R* [2], defined as

$$R\left(\frac{\sigma}{S}\right) = \frac{S(\omega_0)}{B(\omega_0)} \tag{5.1}$$

where σ is the noise standard deviation, S is the threshold level which allows the system to jump ; $S(\omega_0)$ is the amplitude of the output signal power spectrum computed at the forcing signal frequency ω_0; $B(\omega_0)$ is determined by linear interpolation of the discretised $S(\omega)$ after subtracting the point $S(\omega_0)$.

The Signal-to-Noise Ratio, R, experimentally obtained for the analog comparator being considered, is given in Figure 1. The experimental results show the enhancement of the Signal-to-Noise ratio R vs. the noise variance σ, which confirms the existence of stochastic resonance in the analog comparator.

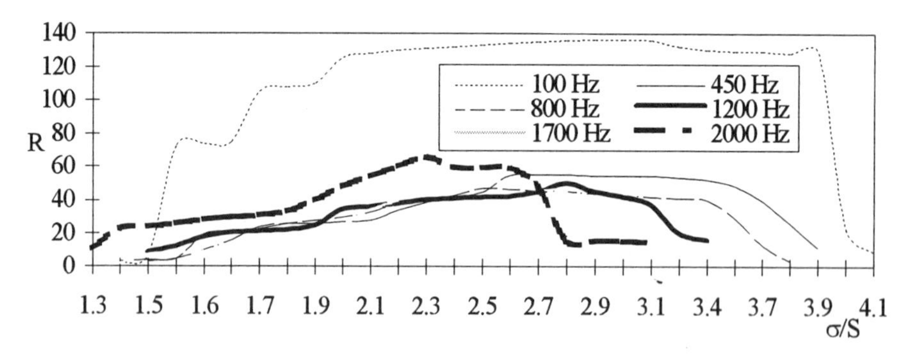

Figure 1. Signal-to-Noise ratio R experimentally obtained for the analog comparator developed, at different forcing signal frequencies.

5.2.1 A VIRTUAL INSTRUMENT FOR REDUCING THRESHOLD ERROR

In order to improve the performance of a generic measurement device, by reducing the threshold error, the standard deviation of the noise generator must be chosen such that the maximum value of R is obtained.

The experimental results regarding stochastic resonance obtained for an electronic comparator show that the standard deviation of the noise, allowing for the persistence of the stochastic resonance phenomenon, depends on both the system threshold and the forcing signal frequency.

A virtual instrument which integrates a traditional analog instrument with an intelligent module for minimisation of the threshold error will now be proposed. Figure 2 shows a functional block scheme of the measuring system developed. It consists of three main sections:

a) evaluation of R vs. σ;
b) evaluation of σ_{opt} value corresponding to the maximum value of R ;

c) a voltage-controlled amplifier (VCA) to drive the noise generator to the σ_{opt} value.

The autotuning control system was created using *LabVIEW*®, while data were gathered using the National Instruments *AT-MIO 16-E-10* data acquisition card.
The optimal value for the noise variance was determined repeatedly; in this way, if a variation in either the threshold value or the input signal frequency occurs they can be compensated by the "autotuning" system.

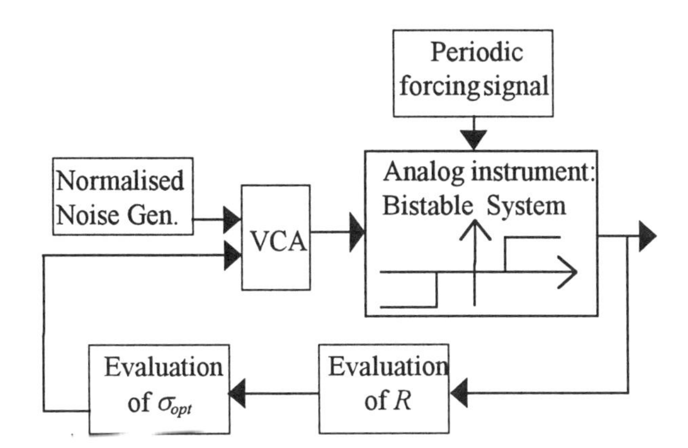

Figure 2. Functional block scheme of the virtual instrument for reducing threshold error.

In Figure 3 the front panel of the proposed virtual instrument is reported. It consists of two parts. The first one deals with acquisition of the system output signal, while the second performs efficient noise variance value tuning, on the basis of the system parameter fluctuations.
An improvement in the performance of the measuring system, consisting of a 30% threshold reduction, was observed
Figure 4 shows the responses of the comparator forced by both a periodic input with an amplitude lower than the threshold and a suitable noise signal, and the same input with an amplitude higher than the threshold.
The system investigated is characterised by a *30 mV* threshold. It can be observed that, in *SR* conditions, signals with an amplitude (*21 mV*) lower than the system threshold cause commutation of the bistable system.
The virtual instrument proposed allows for the automatic selection of the noise variance in order to assert the existence of stochastic resonance in the system that determines a threshold level lower than that in the original device.
The proposed methodology does not depend on the particular analog device considered but can be used to reduce the threshold error of any generic system.

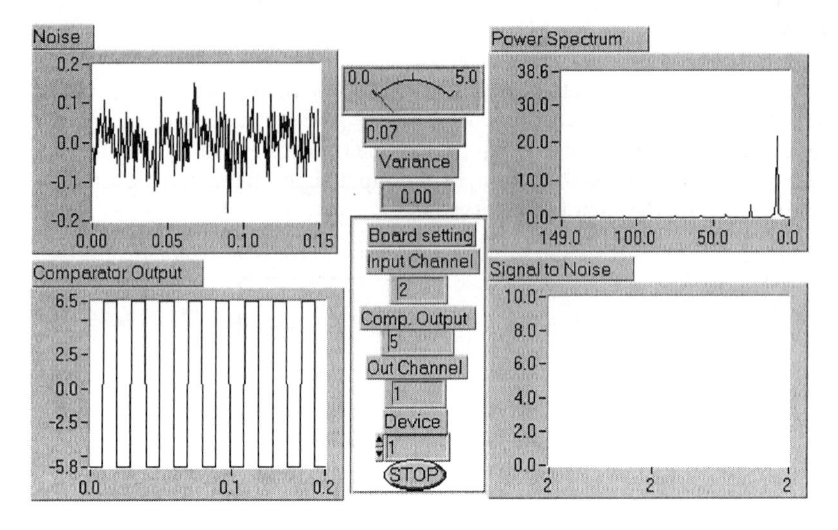

Figure 3. The front panel of the proposed virtual instrument.

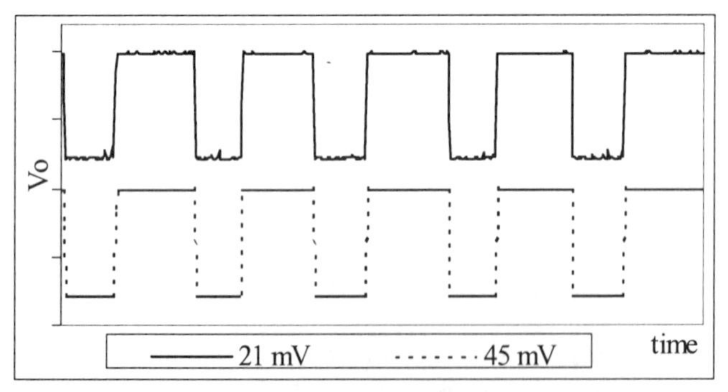

Figure 4. The responses of the comparator, at different periodic input amplitudes.

5.3 Threshold reduction in a Schmitt Trigger

The noise optimisation theory, proposed in Chapter 3, was experimentally validated by applying a noise forcing signal to a bistable Schmitt trigger, forced by a periodical signal, whose amplitude A was lower than the system threshold $S=700\ mV$.

5.3.1 A VIRTUAL DDE-BASED TOOL FOR THE AUTOMATIC EVALUATION OF *F(N)*

The optimisation procedure giving the iso-amplitude map described in Chapter 3 was implemented in *Matlab®*. A virtual instrument, implemented in *LabVIEW®*, uses the *Matlab®* tool both to guarantee a user-friendly interface and to obtain an automatic procedure. The virtual instrument establishes a dialog with the *Matlab®* environment by means of DDE instructions. On the front panel the user can set the frequency range of interest and the range of values to be used for *F(N)*. The form of the expression *F(N)* and the form of the relationship between the noise variance σ_{opt} and the system parameters can be defined by means of a batch script. The instrument transfers all this information to the *Matlab®* function implementing the optimisation algorithm for *F(N)* detection. A schematic representation of the system is given in Figure 5.

It should be borne in mind that the pairs $(A_{min}, \sigma_{opt})^i$ have to be transferred to the optimisation algorithm. In order to perform the task of determining the pairs, a dedicated section of the virtual instrument (schematised on the right-hand side of Figure 5) was developed. Communication with the required instruments is based on the HPIB protocol supported by *LabVIEW®*. By means of this protocol several electronic devices, such as arbitrary function generators and digital scopes, can be physically connected to a PC to be completely controlled. In this way, both the forcing signal parameters and the noise variance can be changed in order to find the experimental pairs $(A_{min}, \sigma_{opt})^i$. A frequency measure is obtained by means of the measuring kit of the digital scope in order to check whether the system commutations occur at the same frequency as the forcing signal. These values, transmitted to the virtual instrument via HPIB, are compared to the forcing signal frequency. If the two values match, the tested couple is saved and is shown to the user. After the couples $(A_{min}, \sigma_{opt})^i$ have been detected the optimisation algorithm can start (left hand side of Figure 5).

The front panel of the virtual instrument developed is presented in Figure 6. The right-hand side of the instrument is dedicated to automatic selection of the couples $(A_{min}, \sigma_{opt})^i$. A waveform graph is used to monitor the output of the device (e.g. the Schmitt trigger) during the selection procedure. The interface allowing data exchange with the optimisation algorithm developed in *Matlab®* is implemented on the left-hand side of the instrument. At the end of the optimisation procedure the *J* index, introduced in Chapter 3, is given in the waveform graph reported on the left-hand side, as a function of the *F(N)* values. The minimum value of this index identifies the *F(N)* to be sought.

5.3.2 EXPERIMENTAL RESULTS

In order to obtain both the minimum values for the signal amplitude and the noise variance, A_{min} and σ_{min}, allowing for system commutations, and the *F(N)* value the above introduced algorithm was used.

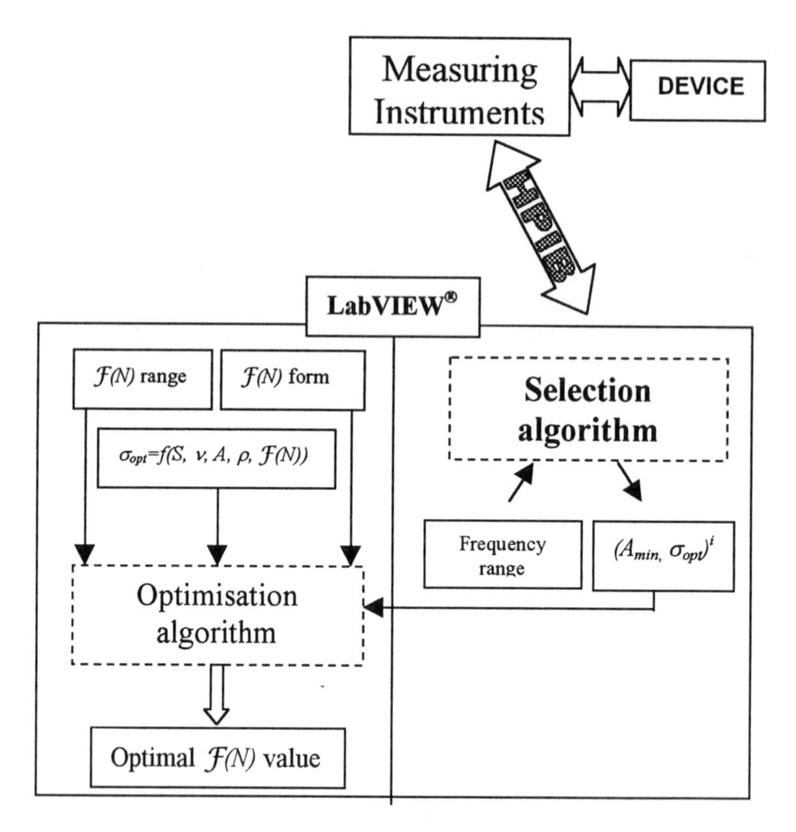

Figure 5. Interaction between the virtual instrument, the computational paradigm and the digital instruments. The virtual instrument and the *Matlab®* environment exchange data via the DDE functions, while communications with the measuring devices take place via the HPIB protocol.

Figure 7 shows a comparison between the experimental and theoretical values of the optimal (minimum) noise signal variance allowing for suitable system commutations. Figure 8 shows the iso-amplitude curves for the experimental trials with the bistable trigger. The black circles show the minimum values of the input signal amplitude allowing for system commutations, and the corresponding minimum noise variance, evaluated at the considered forcing frequency. The thick line divides the map into two zones. For each forcing frequency value, the forcing signal amplitude allowing for system commutation must be sought in the left-hand zone of the map; the corresponding reading in the variance axis will give the minimum noise variance value.

The map in Figure 8 represents a suitable method for parameter tuning, allowing for an improvement in system performance.

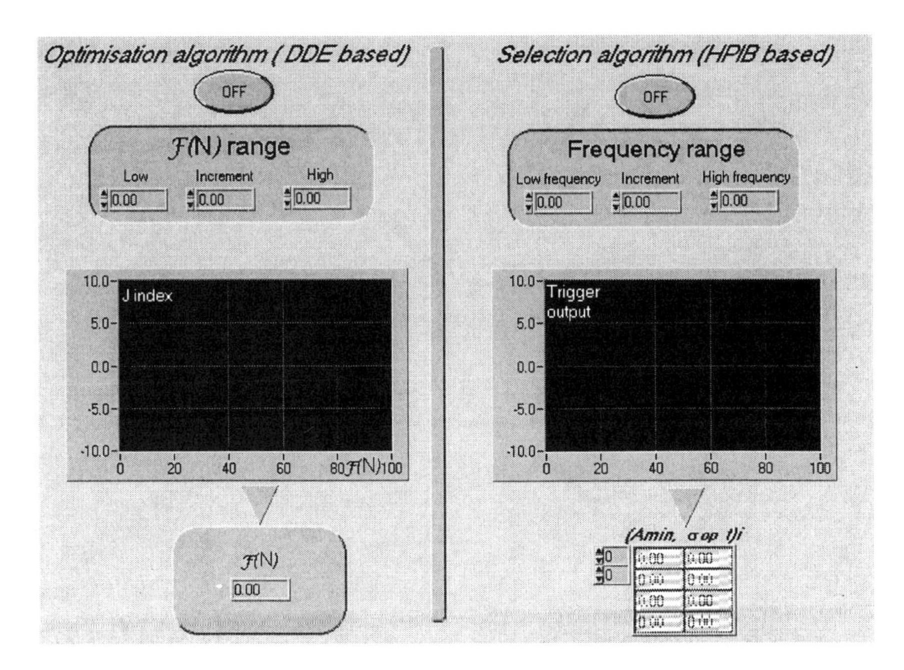

Figure 6. The front panel of the virtual instrument for automatic selection of the couples (A_{min}, σ_{opt}) and the *F(N)* expression. . The right-hand side of the instrument is dedicated to automatic selection of the couples $(A_{min}, \sigma_{opt})^i$. A waveform graph is used to monitor the output of the during the selection procedure. The interface allowing data exchange with the optimisation algorithm developed in *Matlab®* is implemented on the left-hand side of the instrument.

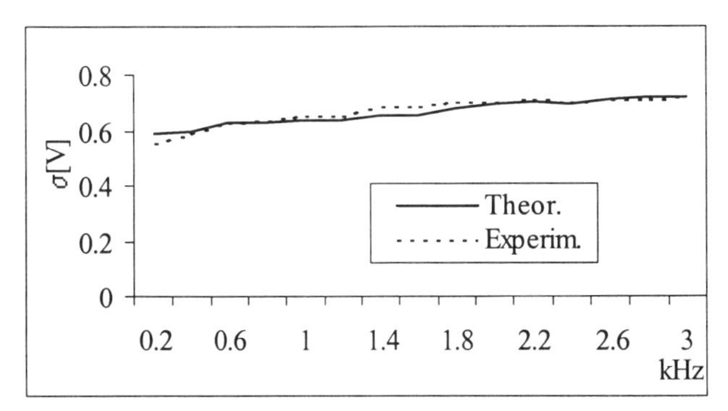

Figure 7. Comparison between the experimental and theoretical values of the optimal (minimum) noise signal variance allowing for suitable system commutations.

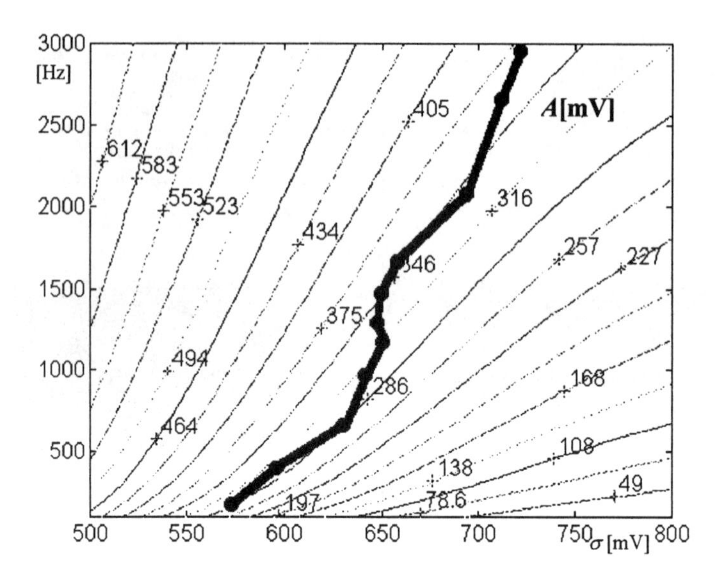

Figure 8. Isoamplitude map. The black circles show the minimum values of the input signal amplitude allowing for system commutations, and the corresponding minimum noise variance, evaluated at the considered forcing frequency. The thick line divides the map into two zones. For each forcing frequency value, the forcing signal amplitude allowing for system commutation must be sought in the left-hand zone of the map; the corresponding reading in the variance axis will give the minimum noise variance value.

In order to check the reliability of the proposed approach, several points of the iso-amplitude map were tested and estimations close to the experimental values were found. For example, if a forcing signal with a *1 kHz* frequency and a *500 mV* amplitude is applied the map gives a $\sigma_{opt}=525\ mV$. To test the validity of this estimation the corresponding system behaviour can be considered. In Figure 9 the trigger output for noise standard deviation values of both $\sigma=515\ mV$ and $\sigma=530\ mV$ is given. In the first case the noise amplitude is insufficient to allow the required commutations, while in the second case the system output commutes at the frequency of the forcing signal.

The possibility of performing *F(N)* detection via a small number of measurements and then estimating the whole iso-amplitude map by a numerical procedure represents the main feature of the proposed approach. As an alternative solution, the whole parameter map should be identified, but a larger number of measurements would be required.

5.3.3 NOISE CONTROL LAW IMPLEMENTATION

In order to realise an analog device allowing suitable noise variance control, a simple form of the control law (Equation 3.21 in Chapter 3) must be sought for. The simplest form is the polynomial one

$$\sigma=\lambda_1 A+\lambda_2 A^2+\lambda_3 A^3+.......+\lambda_m v+\lambda_{m+1} v^2+\lambda_{m+2} v^3+......+\lambda_n \tag{5.2}$$

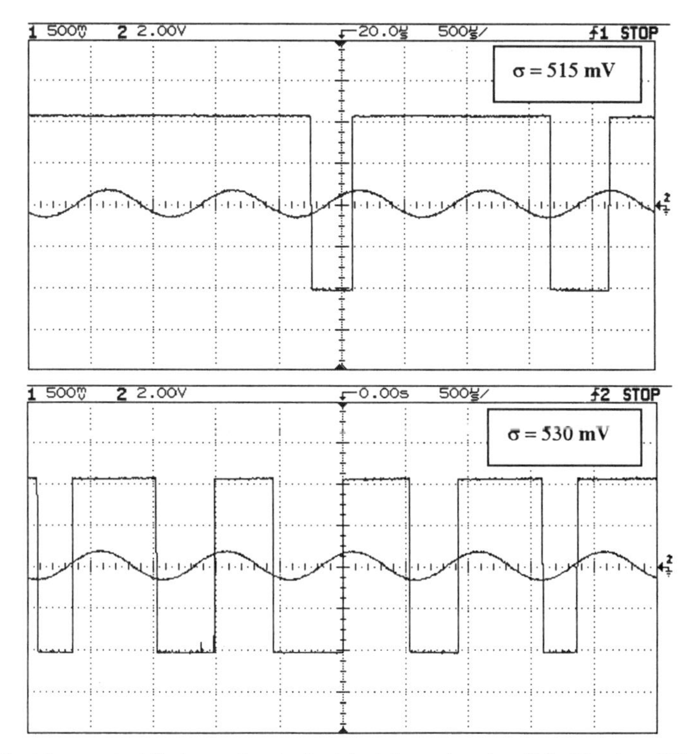

Figure 9. The trigger output for two noise standard deviation values (σ=*515 mV* and σ=*530 mV*), due to a forcing signal with a frequency of 1KHz.

For evaluation of the λ_i parameters the pattern given by the iso-amplitude map must be used. Several attempts have been made in order to find a suitable form of (5.2) and, for the case being discussed, the following relation gives the best system performance

$$\sigma=\lambda_1 A+\lambda_2 v+\lambda_3 \tag{5.3}$$

In order to estimate the model parameters, an LMS-based algorithm was implemented and the model parameters were identified using the analytically-generated data set shown in Figure 8.

Figure 10 shows a comparison between the original and the estimated data, using model (5.3). In Figure 11 the data density for the estimation residuals is also plotted. By using a very simple form of relationship (5.2) a good agreement between the original and the estimated data has been obtained.

Figure 12 is a schematic representation of the whole system. Sections **A** and **B** are used for frequency-to-voltage conversion and detection of the amplitude of the periodical signal, respectively.

This information is sent to the device **C** implementing model (5.2). The parameter to be controlled is the voltage V_G to be sent to the Voltage Controlled Amplifier allowing noise amplitude modulation.

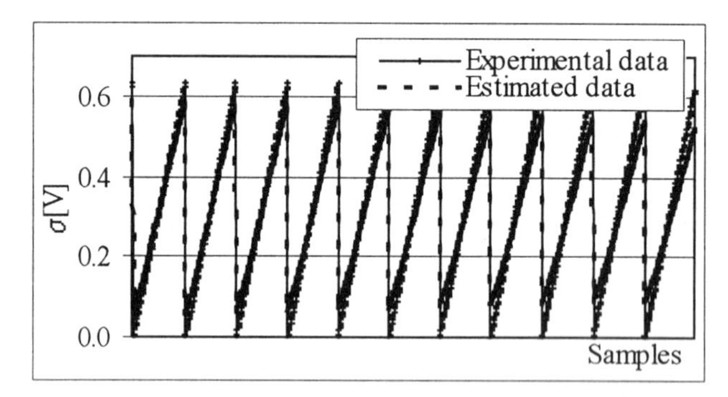

Figure 10. Comparison between the original standard deviation and the estimated one, (Equation 5.3).

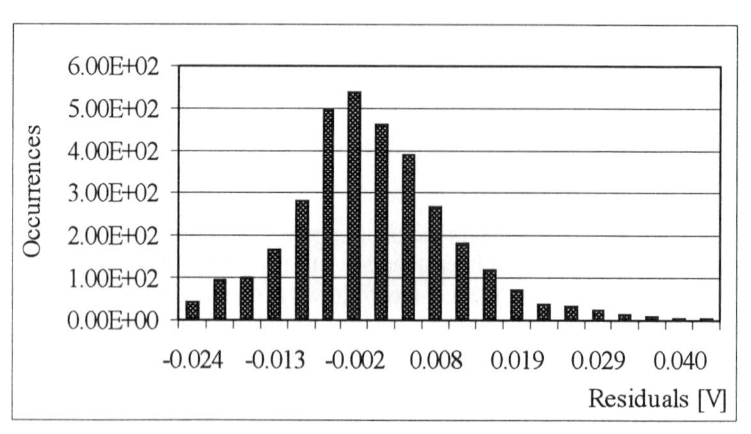

Figure 11. The data density plot for the residuals of the estimation shown in Figure 11.

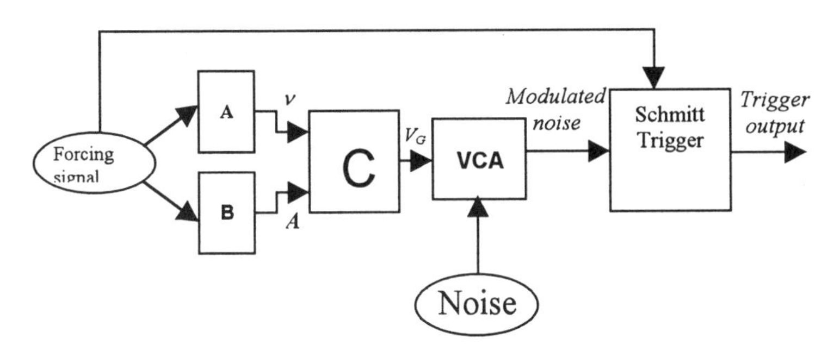

Figure 12. The schematic representation of the analog noise variance control system.

5.3.4 A VIRTUAL INSTRUMENT FOR OPTIMAL NOISE VARIANCE DETECTION

The control block **C** (or model 5.3) was first implemented by means of a virtual instrument realised using *Labview®* software. The *AT-MIO 16-E10* board was used to interface the software control and the analog device. Information about the frequency and the amplitude of forcing signal must, in fact, be given as input data to the virtual instrument. In order to extract this information from the input signal, both a frequency-to-voltage converter and a peak follower device were used These electronic devices adopted were characterised and their input-output characteristics included in model (5.3).Of course, the virtual instrument must return a voltage proportional to the optimal noise variance value. This signal was used to modulate the noise signal, by an analog mixer.

In Figure 13 the front panel of the proposed instrument is shown for generic values of the parameters processed. The right-hand side section of the panel, *A*, is dedicated to representation of the top-down structure already shown in Chapter 3.

The top part of section *B* is used to give information about the acquired forcing system parameters. A real-time procedure allows detection of the optimal noise variance value, on the basis of the *A* and *v* values and the identified model. The detected variance value is continuously shown in the bottom left part of the panel.

As can be observed in Figure 13, the parameter values match the theoretical estimation shown in Figure 6.

Section C of the panel is used to perform manual selection of the noise variance value. By acting on the switch the noise amplitude auto-searching feature of the instrument is disabled and the horizontal slide allowing manual noise amplitude control is activated.

Good performance was achieved with the proposed instrument, even in a periodic signal frequency range overlapping with the one used during estimation of the model (5.3).

5.3.5 ANALOG IMPLEMENTATION OF THE NOISE VARIANCE TUNING SYSTEM

The analog realisation of model (5.3) is the last step in the procedure. In this case the information about the forcing signal amplitude and frequency must be sent to the analog device implementing the model (5.3), as shown in Figure 14. The parameter to be controlled is the voltage V_G to be sent to the Voltage Controlled Amplifier allowing noise amplitude modulation. By using the manual switch **S** the control input of the VCA can be switched to manual noise amplitude modulation, as implemented by section **D**, to the analog model **C** and to the null noise mode. The experimental set-up of the proposed device is shown in Figure 15.
Good performance was obtained with the system in both the frequency and amplitude ranges considered.

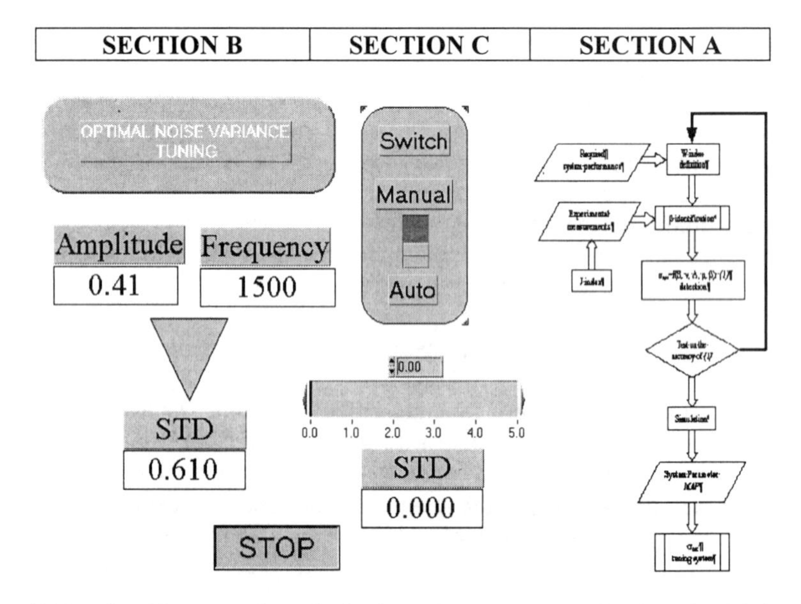

Figure 13. The virtual instrument for optimal noise variance parameter estimation. The right-hand side section of the panel, A, is dedicated to representation of the top-down structure already shown in Chapter 3. The top part of section B is used to give information about the acquired forcing system parameters. A real-time procedure allows detection of the optimal noise variance value, on the basis of the A and ν values and the identified model. The detected variance value is continuously shown in the bottom left part of the panel. Section C of the panel is used to perform manual selection of the noise variance value.

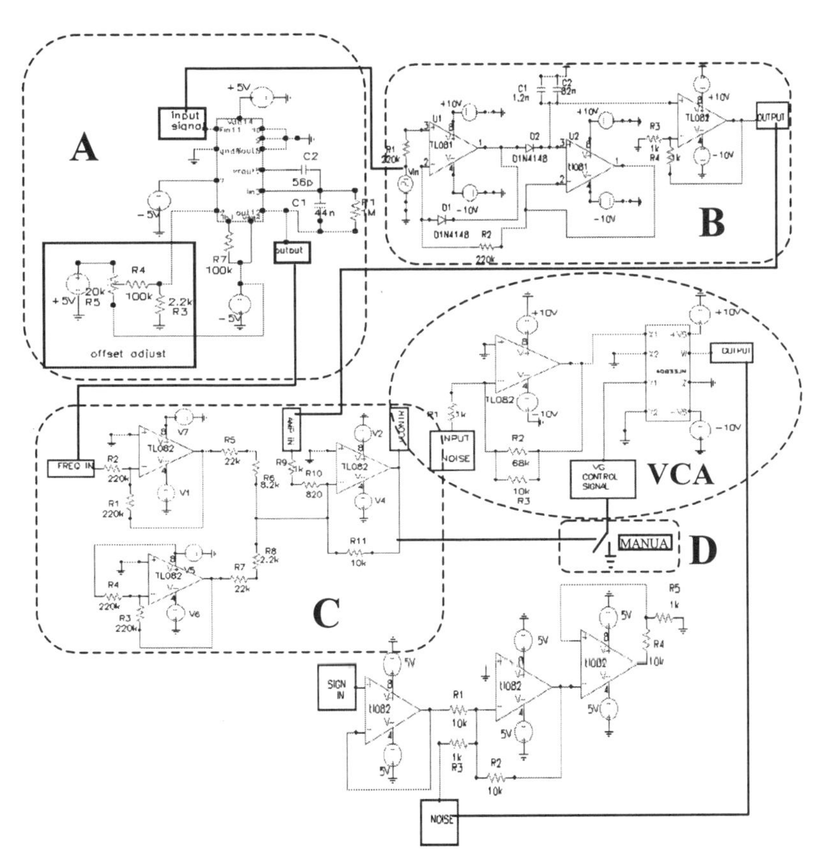

Figure 14. Schematic representation of the noise variance control device.

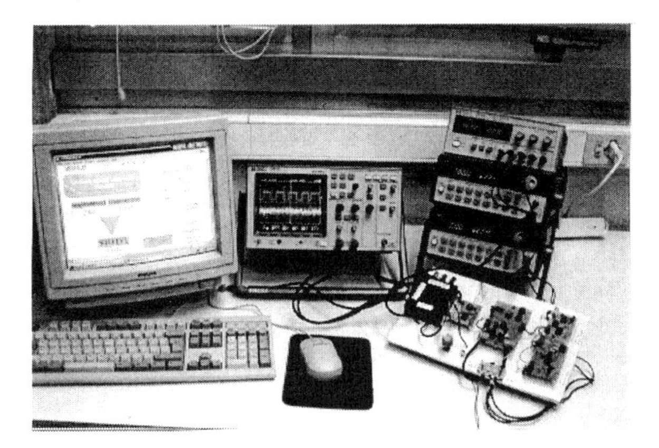

Figure 15. Experimental set-up of the noise tuning system for the Schmitt Trigger.

5.4 Damped travelling waves in non autonomous *CNN*s by noise modulation

CNN based circuits are well known distributed circuit architectures where several interesting phenomena have been widely studied. In particular, a lot of contributions have been recently reported in literature, showing that *CNN* based circuits are very suitable to generate complex phenomena like autowaves, spiral waves, and to model several spatio-temporal phenomena [15]. So far, the most exploited *CNN* paradigm to this purpose has been the autonomous one, in which circuits without input signals are taken into account. Moreover, great attention has been paid to the active waves phenomena, in which waves are self sustained [16].
In this section, the case of damped travelling waves in non autonomous *CNN*s is discussed [17]. A forced wavefront imposed on an edge of the considered bidimensional structure, damped due to the nature of the circuits, is able to propagate along the structure if a small input signal with random spatio-temporal characteristics is imposed at the input of each cell.
The adopted Reaction-Diffusion *CNN* (*RD-CNN*) has been designed in two steps. First, the single cell model is derived in order to exhibit a non-linear, non autonomous oscillation. In particular, the following second-order damped system has been considered

$$\begin{aligned} \dot{x}_{1,i,j} &= -x_{1,i,j} - y_{2,i,j} + u \\ \dot{x}_{2,i,j} &= -x_{2,i,j} + y_{1,i,j} + y_{2,i,j} \\ y_{k,i,j} &= 0.5(|x_{k,i,j}+1| - |x_{k,i,j}-1|) \end{aligned} \tag{5.4}$$

where u is the input signal of each cell, $i=0,1,\dots,M-1;\ j=0,1,\dots,N-1;\ k=1,2$.

The second step consists in coupling the *CNN* cells via diffusion coefficients which allow non-isotropic propagation of a forcing signal imposed to a *CNN* edge. Therefore, it is possible to derive the following RD-*CNN* with constant templates

$$\dot{x}_{ij} = -x_{ij} + A * y_{ij} + B * u_{ij} + I \tag{5.5}$$

where $x_{ij}=[x_{1;i,j}\ \ x_{2;i,j}]^T$, $y_{ij}=[y_{1;i,j}\ \ y_{2;i,j}]^T$ and $u_{ij}=[u_{1;i,j}\ \ u_{2;i,j}]^T$ are the state, the output and the input of the *CNN* respectively; while A, B and I are the so-called feedback, control and bias templates, respectively, and T is the transpose operator.

The templates are

$$A = \begin{pmatrix} A_{11} & A_{12} \\ A_{21} & A_{22} \end{pmatrix} \qquad A_{11} = \begin{pmatrix} 0 & 2 & 0 \\ 0.1 & -0.2 & 0.1 \\ 0 & -2 & 0 \end{pmatrix}$$

$$I = 0 \qquad A_{22} = \begin{pmatrix} 0 & 0 & 0 \\ 0 & 1 & 0 \\ 0 & 0 & 0 \end{pmatrix}$$

$$B = \begin{pmatrix} B_{11} & 0 \\ 0 & 0 \end{pmatrix} \qquad A_{12} = -A_{21} = \begin{pmatrix} 0 & 0 & 0 \\ 0 & -1 & 0 \\ 0 & 0 & 0 \end{pmatrix}$$

$$B_{11} = \begin{pmatrix} 0 & 0 & 0 \\ 0 & 1 & 0 \\ 0 & 0 & 0 \end{pmatrix}$$

The input signal u consists of a sinusoidal signal applied to the upper edge of the *CNN* first layer. Due to the natural passivity of the system, the propagation of the wave imposed by the term u is damped within a few rows of the *CNN*, as it is shown in Figure16.

A spatio-temporal noise $s_{i,j}(t)$ has been generated by using a *uniform* distribution and has been applied as input to each cell in equation (5.4). The effect of imposing a suitable spatio-temporal noise level $s_{i,j}(t)$, supporting the wavefront transmission, is shown in Figure 17. The goodness of the results depends also on the spatial degree of correlation of the noise, which is determined by the number of *CNN* cells affected by the same noise function. In the studied case, it has been imposed the same random noise on patches of 4X4 cells, that is comparable to the length scale of the imposed travelling wave width, which assures the best propagation.

In Figure 18, it is shown that a under-threshold noise level $s_{i,j}(t)/2$ applied to the *CNN* structure does not support the wavefront propagation.

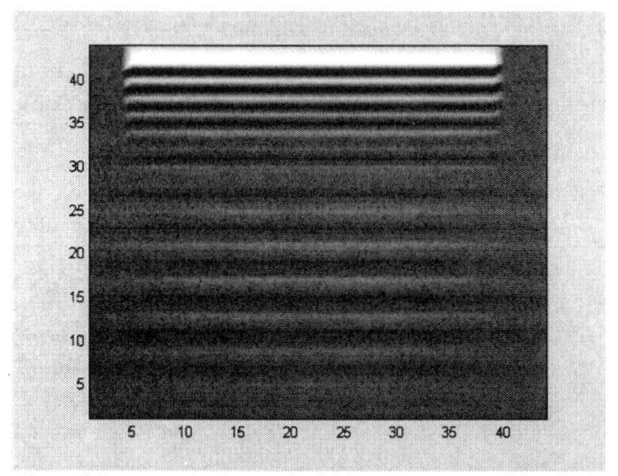

Figure 16. The propagation of the wave is damped within a few rows of the *CNN*.

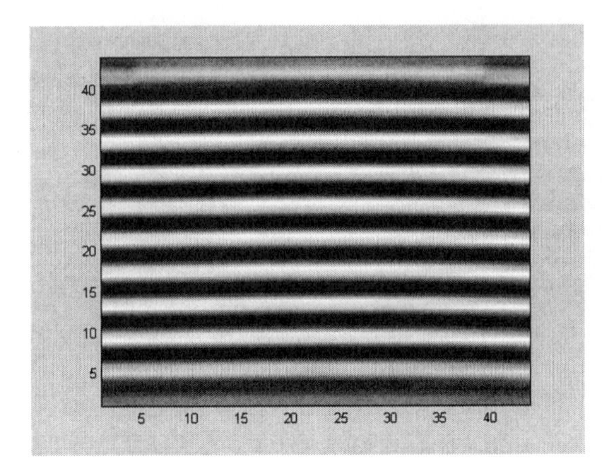

Figure 17. The effect of imposing a suitable spatio-temporal noise level.

In Figures 19, the effect of increasing the noise level up to $1.5s_{i,j}(t)$ and $5s_{i,j}(t)$ are respectively shown. Thus, it is clearly shown that there exists an optimal value for the noise level, beyond which propagation defects arise.
These considerations denote the peculiarity of the stochastic resonance phenomenon in non autonomous *CNN*s.

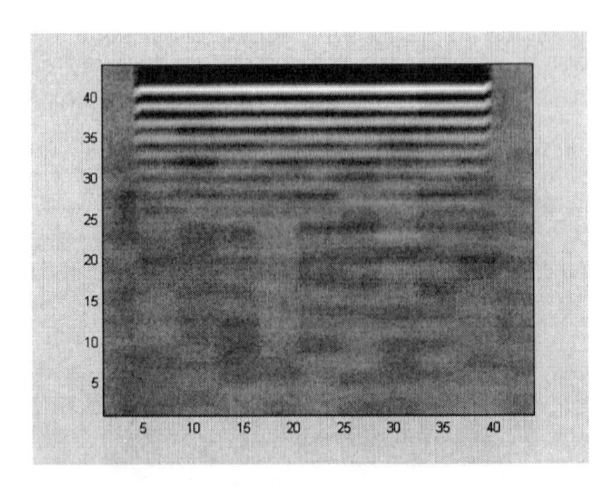

Figure 18. Under-threshold noise level si,j(t)/2 applied to the *CNN* structure does not support the wavefront propagation.

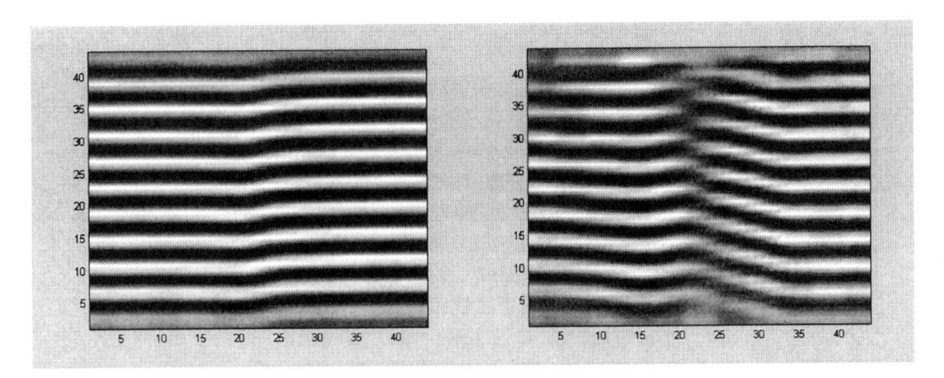

Figure 19. The effect of increasing the noise level up to $1.5s_{i,j}(t)$and $5s_{i,j}(t)$.

5.5 Threshold error reduction in an optical transmission system by suitable noise modulation

In this section an application of the linearisation procedure introduced in Chapters 2 and 3 is proposed. It should be reached that the proposed approach allows minimisation of non-linear system behaviour by using the linearising capability of noise-based techniques [10-13]. In particular, an improvement can be obtained in the features of several linear *measuring devices* (such as optical sensors, environmental sensors, proximity sensors, etc.) [9,14].

Let us consider a generic contactless transmission measuring device. If the amplitude of the signal to be transferred is gradually increased from zero there will be some minimum value below which the signal cannot be conveyed. This minimum value defines the threshold of the device [14], or the threshold error, which is one of the relevant errors in measuring devices.

In this section an electro-optical device is considered. Electro-optical devices exhibit very interesting properties, but threshold error represents a serious problem in many applications for this class of measuring devices and it should be kept as low as possible for optimal system performance [14].

As a first experimental validation of the theory developed, a coupled phototransistor detector/diode emitter, shown schematically in Figure 20, is investigated. The forcing signal is a sinusoidal wave with a frequency f=1.0 Hz. The device threshold was experimentally evaluated and the value S=1.0 V was obtained. With an input amplitude lower than 1.0 V, the received signal amplitude is lower than the threshold value of the sensor and the output signal cannot be correctly detected. Hence, threshold error may cause a serious lack of information.

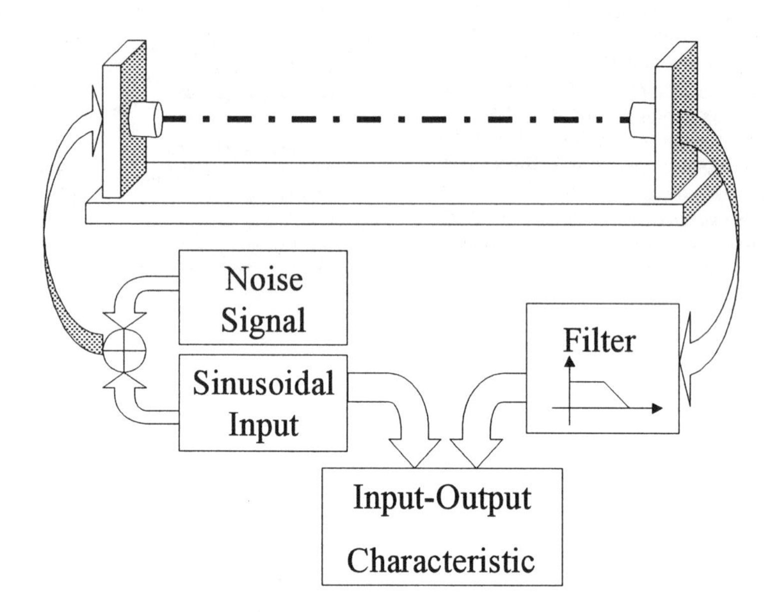

Figure 20. Schematic representation of the coupled phototransistor detector/diode emitter system.

Both the input and the output signals are shown in Figure 21. Comparison between the measuring device output and the forcing signal shows that the device cannot work properly with such low input signal values.

In order to obtain optimal device performance the threshold error should be eliminated. According to the theory introduced in Chapter 3, a noise signal with $\sigma_{opt}=S=1.0\ V$ is required. The noise signal to be added to the periodic input is generated by using the Hewlett Packard HP33120A signal generator. The device output is processed by using a first-order active low-pass filter. The results obtained are shown in Figure 22.

In Figure 23 the input-output characteristics of the device for different noise standard deviation values, σ, are compared. As already mentioned it can be observed that optimal linearisation is reached when $\sigma=\sigma_2=S$.

A further experimental validation of the result obtained is possible by using an experimental index J. Considering that a linear device characteristic is required, the following performance index is defined

$$J=\frac{1}{N}\sqrt{\sum_{i=1}^{N}\left(\underline{\xi_i}-\xi_i\right)^2} \tag{5.6}$$

where ξ_i and ξ_i represent points of the theoretical linear characteristic and the measured one respectively.

The index J is used to represent the effect of the noise signal variance on the linearity of the system characteristic in the extreme region of the working range. Figure 24 shows the index J as a function of the noise variance. According to the above considerations, the noise standard deviation value $\sigma_{opt}=S=1.0\ V$ assures the minimum value of J.

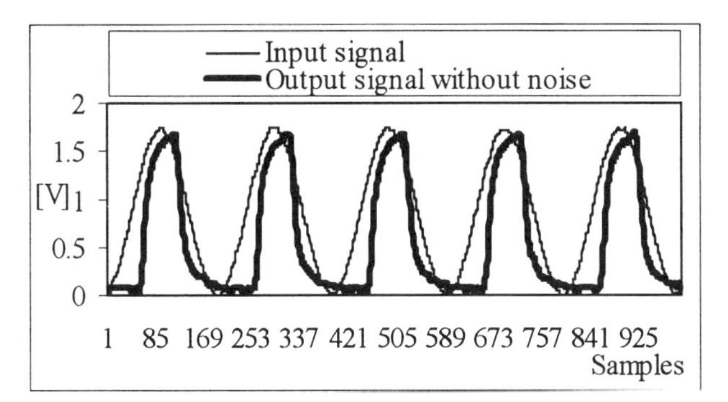

Figure 21. Input and output signals of the transducer considered. The response of the measurement device, compared to the forcing signal, shows that a low input signal cannot be correctly detected.

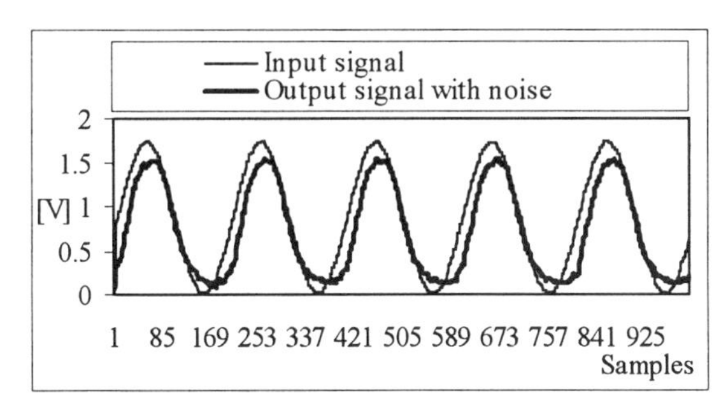

Figure 22. Input and output signals after the linearising procedure has been applied.

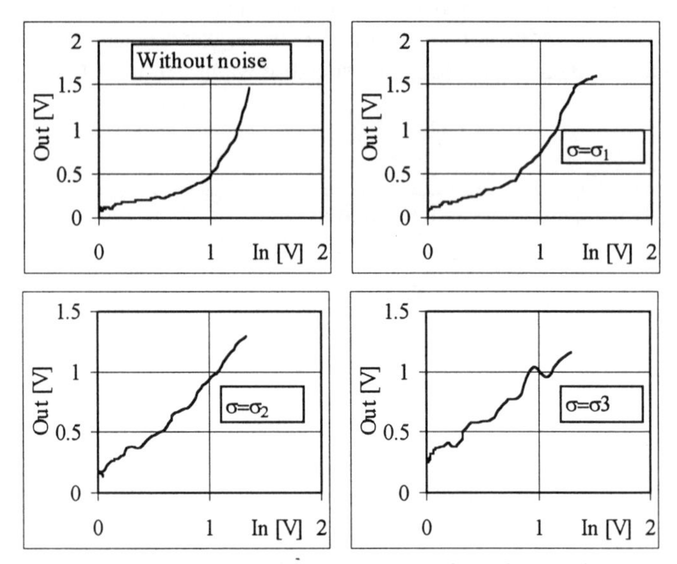

Figure 23. Input-output characteristic of the device without noise action and for three noise standard deviation values. It can be observed that optimal linearisation is reached when $\sigma = \sigma_2 = S$.

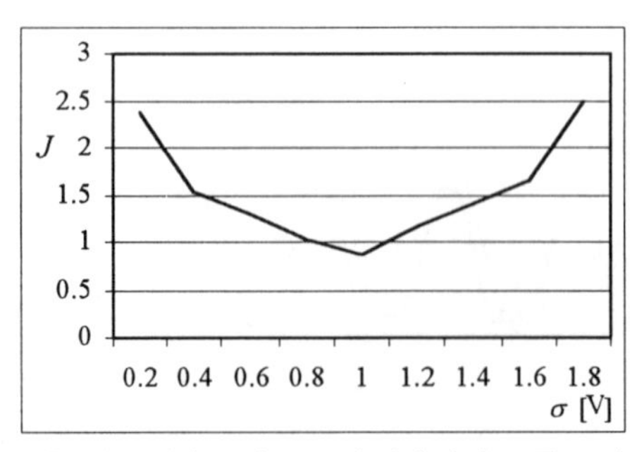

Figure 24. The index J as a function of the noise standard deviation. The noise standard deviation value $\sigma_{opt} = S = 1.0\ V$ assures the minimum value of J.

5.6 A noise acted IR sensor for displacement measurement

Sensors for distance measurement are among the most widespread categories of sensors in the field of commercial applications, and a variety of sensors of this kind are manufactured and marketed by a large number of firms.
As is to be expected, quality and performance on the one hand and economic interests on the other are the main factors regulating the market for these devices. It is therefore necessary to develop new techniques to improve the specifications of low-cost devices. Noise-added techniques can make a great contribution towards achieving this.
Even if the application of these techniques often implies the use of conditioning sections, current costs in electronics amply justify their use if they are capable of making a significant improvement in the system. A good example is an infrared (*IR*) optical sensor used for distance measurement. The advantages of this kind of sensor are its flexible radiance degree and the absence of "contact". It depends greatly, however, on the level of lighting in the environment in which it operates and the nature of the target: surfaces of different colours and with different reflection indexes vary the intensity of the wave received [14].
IR optical sensors generally comprise an emitter and a photo-detector (photodiode or phototransistor). By forcing the emitter with a periodical signal or a continuous voltage, the intensity of the radiation generated is proportional to the amplitude of the forced signal. The wave reflected by the target hits the receiving phototransistor and its intensity basically depends on the distance of the target, even though it may at times be affected by the operating environment. By measuring the intensity of the wave received, it is therefore possible to calculate the distance of the target.
The measurement system used is made up of a trolley which moves freely along a track, a target mounted on the trolley and an *IR* sensor. A reference index on the target indicates its distance from the sensor.
The transmission and reception sensors are mounted on one of the ends of the track and lie on a surface parallel to the target. A scheme of the measurement system is shown in Figure 25.
The target and the sensors were mounted in such a way as to guarantee that the IR radiation and the target were orthogonal; the track ensures that the movement always follows the same trajectory.
The modulating signal is sinusoidal with a frequency of 20 kHz and is forced in a voltage-current converter so as to drive the emitter photodiode correctly.
By inserting the foto-transistor into an amplification network, as shown in Figure 26, the output voltage obtained is equal to the product of the input power and the feedback resistance, proportional to the weak emitter current.
The amplification stage output voltage is sent to a 3rd order band-pass filter which can be implemented with a network like the one shown in Figure 27. The task of the filter is to reduce the effect of interfering signals outside the 5 kHz - 30 kHz band that may affect correct functioning of the sensor, for instance discharge from

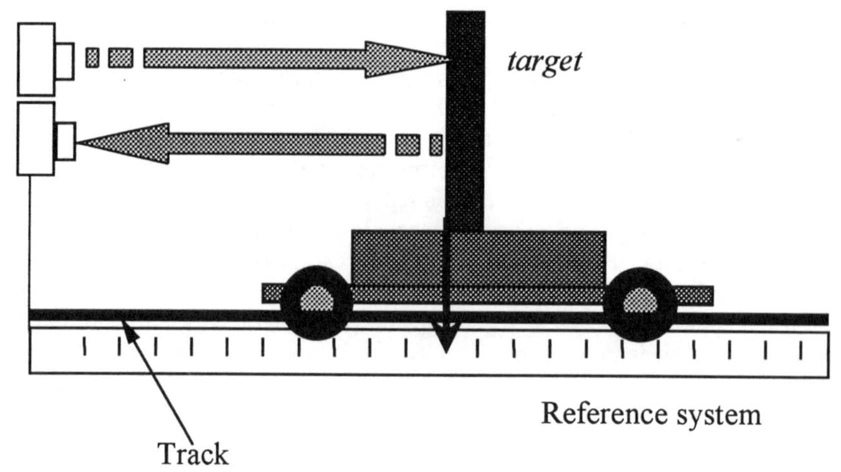

Figure 25. Scheme of the measurement system.

neon light. Finally, the peak detector shown in Figure 28 follows the variations in the amplitude of the signal due to variations in the distance between the target and the sensor. A basic scheme outlining the functioning of the measurement system as a whole is given in Figure 29.

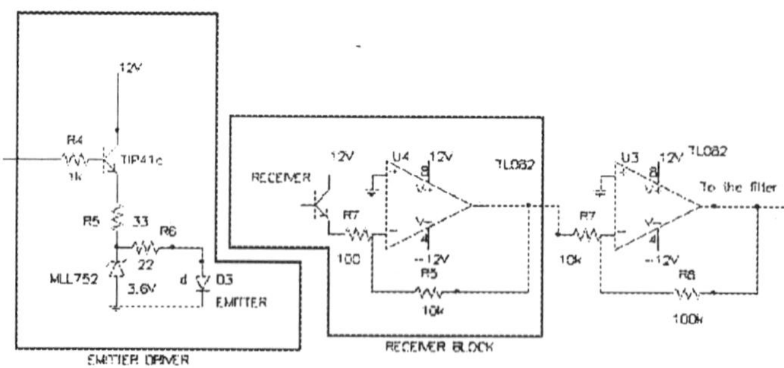

Figure 26. Sensor conditioning circuit.

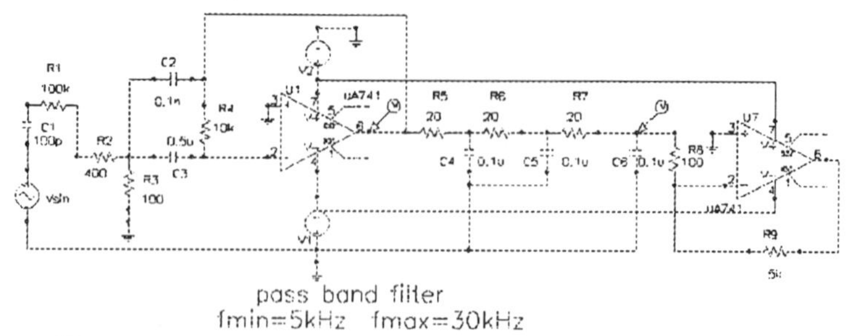

Figure 27. 3rd Pass-band filter; the task of the filter is to reduce the effect of interfering signals outside the 5 kHz - 30 kHz band that may affect correct functioning of the sensor.

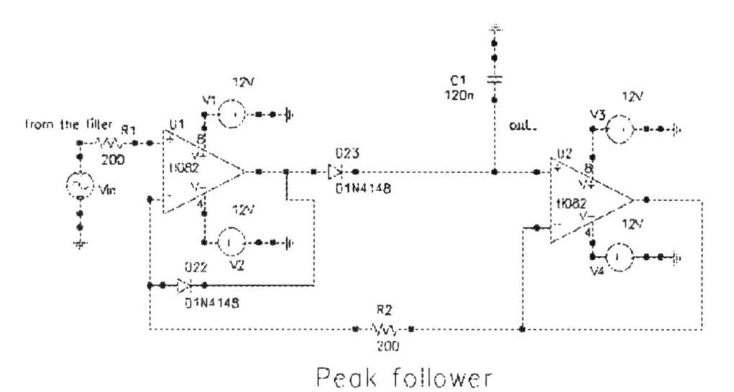

Figure 28. Peak follower.

A number researchers have studied this kind of devices and theoretical works exist that suggest the transfer function characteristic of optical displacement sensors. In [18] the following transfer function is suggested

$$P = \frac{2P_E}{\left(1 + \frac{2h}{z_a}\right)^2} e^{-\frac{8}{\left(1+\frac{2h}{z_a}\right)^2}} \tag{5.7}$$

where

P	is the power collected by the receiver
P_E	is the optical power emitted by the transmitting fiber
H	is the target distance
z_a	is a parameter depending on the receiver/transmitter geometry.

Based on the slight differences between the structure proposed in literature and the devices considered in the experimental set-up a *'grey-box'* identification approach was used to characterise the system.
The following expression was, in fact, supposed for the real device

$$P = \frac{2\alpha}{(1+\beta h)^2} e^{-\frac{\gamma}{(1+\beta h)^2}} + \delta h + \xi \tag{5.8}$$

where, the parameters $\alpha, \beta, \gamma, \delta$ and ξ have been determined by using a Nealder-Mead type search algorithm. In particular, the linear part of expression (5.8) has been introduced to model the linearising effect of noise added approach.

To characterise the measurement system an opaque white reflecting surface orthogonal to the beam emitted by the transmitter was used.
The experimental characteristic, obtained using a sinusoidal forcing signal with an amplitude of 6 Vpp and a frequency of 20 kHz, is shown in Figure 30.
In these conditions a maximum distance of 18 cm was reached, which represents the physical threshold, S_f, of the system and corresponds to an output voltage of 30 mV. This is the threshold voltage of the device and can be indicated with Vs. At distances of over 18 cm the output voltage recorded remained constant at 30 mV.

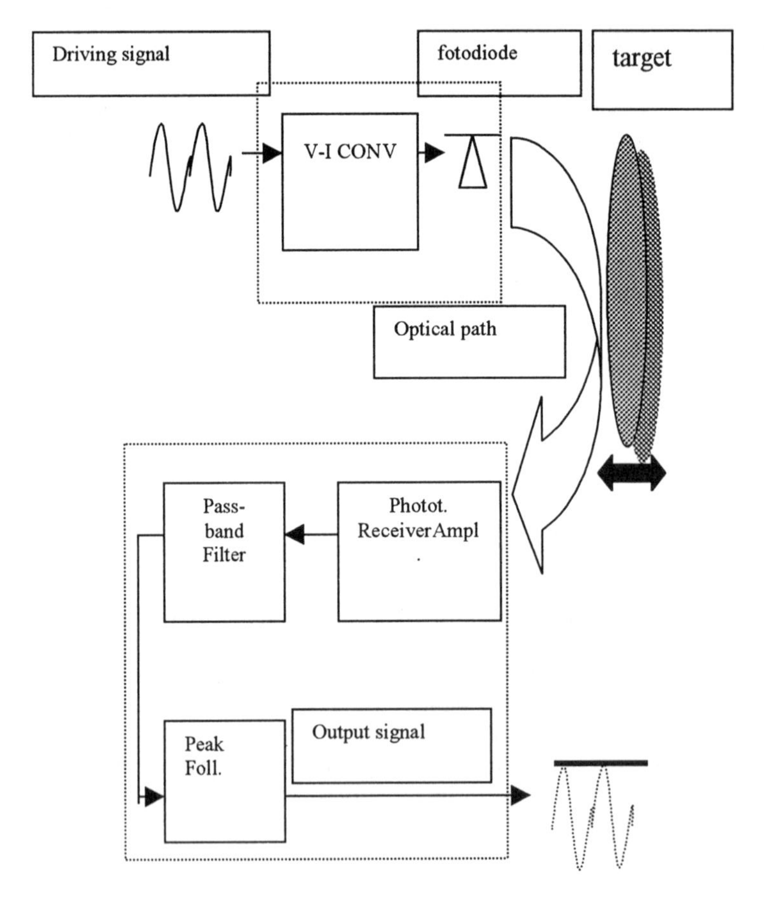

Figure 29. Basic scheme outlining the functioning of the measurement system.

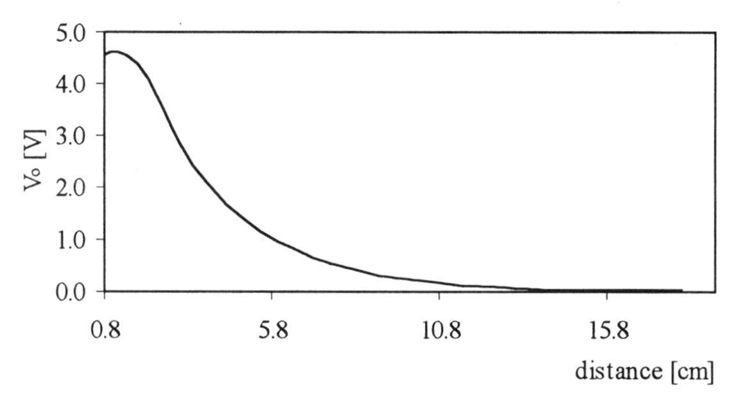

Figure 30. Calibration curve of the IR sensor.

5.6.1 THRESHOLD ERROR REDUCTION VIA NOISE MODULATION

In this section we will assess the behaviour of the *IR* sensor when a *Gaussian* or *uniform* noise is added to the sinusoidal forcing signal.

It is important to be able to determine the optimal noise standard deviation, both for noise generated using a *CNN* and for noise generated by numerical laboratory generators. It is recalled, in fact, that noise with an excessive amplitude would considerably reduce the signal-to-noise-ratio and this would inevitably lead to a degradation in the accuracy.

Initially, to analyse the behaviour of the sensor when modulated by *Gaussian* noise, a lab generator was used. To determine the sensor calibration curves for different noise variances, measurements were made covering the whole field of measurement. The measurements were taken using a sinusoidal input signal with the same frequency and amplitude used to characterise the sensor with no added noise. The physical measurement system is still the one shown in Figure 25 and the reflecting target still has an opaque white surface. This ensures the same functioning conditions and makes it possible to compare the performance of the device with and without added noise. The sensor was then characterised with increasing *Gaussian* noise standard deviation values.

Some calibration curves are shown in Figure 31, which also gives the characteristic obtained without added noise, for purposes of comparison.

When stochastic signals whose standard deviation is lower than the values shown in Figure 31 are used, the system behaves more or less as it does with no added noise. When, on the other hand, the standard deviation is higher, it causes excessive uncertainty in the measurements (introducing the *SNR* problems mentioned previously) and saturation of the output signal.

As can be seen from the graph, with stochastic modulation the sensor's sensitivity in the outlying area of the field of measurement increases as compared with its nominal capacity. So, with optimal standard deviation values the performance of

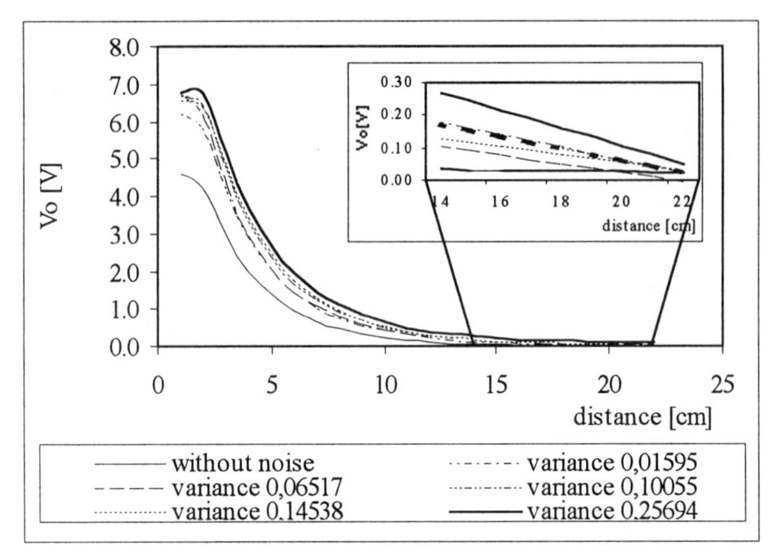

Figure 31. Calibration curves at different values of the noise variance [V^2] when a *Gaussian* noise signal, generated by a digital waveform generator, is used.

the measuring device, in terms of the maximum distance that can be measured, improves considerably. More specifically, optimal performance is achieved with σ=0.5069V. In this case the field of measurement grows from 18 to 22 cm and provides an increase in sensitivity of about 10% in the outlying area of the field of measurement.

On the basis of the theoretical considerations made in Chapters 2 and 3, it emerged that the optimal noise standard deviation to linearise the system was given by the relation $\sigma_G = S$. In the system being considered the receiver threshold is 6 mV. With an initial threshold voltage of V_S=30mV, in fact, it is necessary to consider the gain, G_0, introduced by the post-processing sections, which is equal to 5.

It is also important to consider that the noise signal reaching the receiver (phototransistor) is attenuated, as compared with the signal emitted, by the transmission medium. Analysis of this attenuation mechanism, with a distance of 18 cm between the sensor and the target, gave the results shown in Figure 32. As can be seen, an emitted noise signal with a standard deviation of 0.5069V corresponds to a received signal with a standard deviation of 6.5mV. This result confirms theoretical prediction of the optimal noise variance.

The calibration curves for the sensor forced by a uniformly distributed stochastic component are shown in Figure 33. Considerations similar to those made in the previous case suggested investigating the range of noise amplitudes shown in the figure. More specifically, when a *uniform* noise signal with a standard deviation

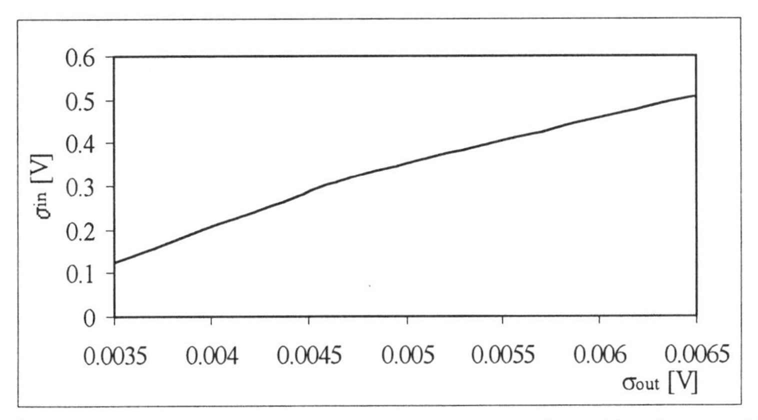

Figure 32. Analysis of the *Gaussian* noise level attenuation mechanism, with a distance of 18 cm between the sensor and the target.

of 0.7437 V is added, the maximum distance measurable is 24 cm and a 12% increase in sensitivity in the outlying area of the field of measurement.
Here again theory gives an optimal value of $\sigma_{opt}=V_s/\sqrt{3}$. Analysis of the noise attenuation shown in Figure 34 shows that an emitted noise with a standard deviation of 0.35 V corresponds to a received noise with a standard deviation of 3.9 mV. This is comparable to the value obtained when the receiver threshold value (6 mV), along with the gain introduced by the post-processing sections (G_0=5), is divided by $\sqrt{3}$.
The calibration curves obtained using non-linear networks (*CNN*s) as the *Gaussian* and *uniform* stochastic sources are shown in Figures 35 and 36 respectively. As before, there is an evident improvement in the performance of the device.
The calibration curves obtained in the previous cases are reported in Figures 37 for purposes of comparison.
Besides the field of measurement and sensitivity, analysis of the performance of a measurement system also has to take the calibration diagram into account. The uncertainty associated with measurements is, in fact, a discriminating factor in assessing the performance of any type of measuring device. Figures 38-40 give the calibration diagrams for all the cases examined. The 3σ limits imprecision is used. Comparison between the functioning of the sensor with and without added noise gives comparable uncertainty bands. Figures 41 illustrates the experimental set-up of the device proposed.

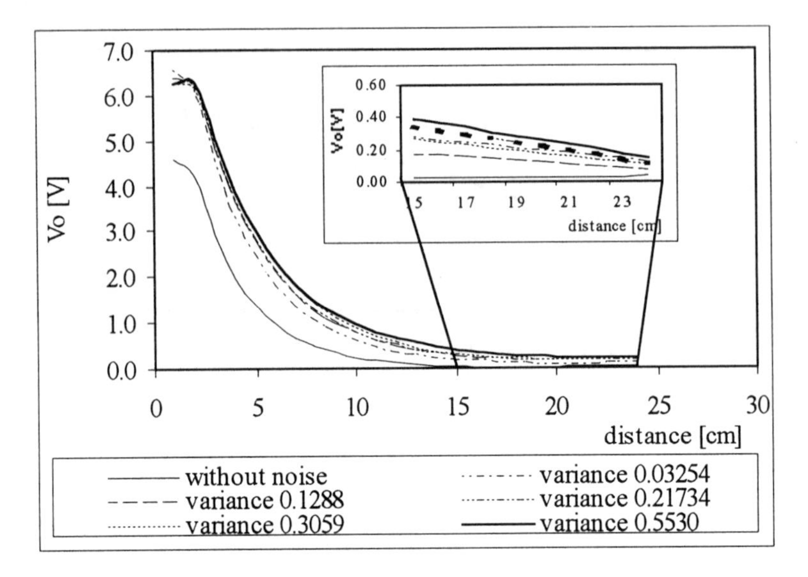

Figure 33. Calibration curves at different values of the noise variance [V^2] when a *uniform* noise signal, generated by a digital waveform generator, is used.

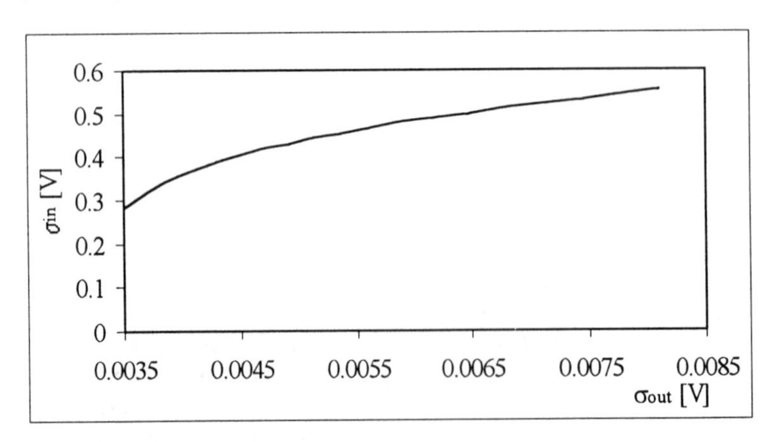

Figure 34. Analysis of the *uniform* noise level attenuation mechanism, with a distance of 18 cm between the sensor and the target.

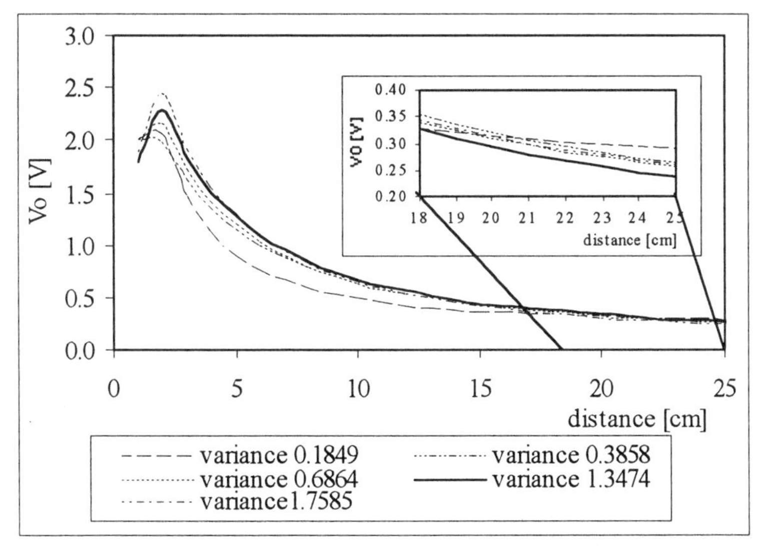

Figure 35. Calibration curves at different values of the noise variance [V^2] when a *Gaussian* noise signal, generated by a *CNN*, is used.

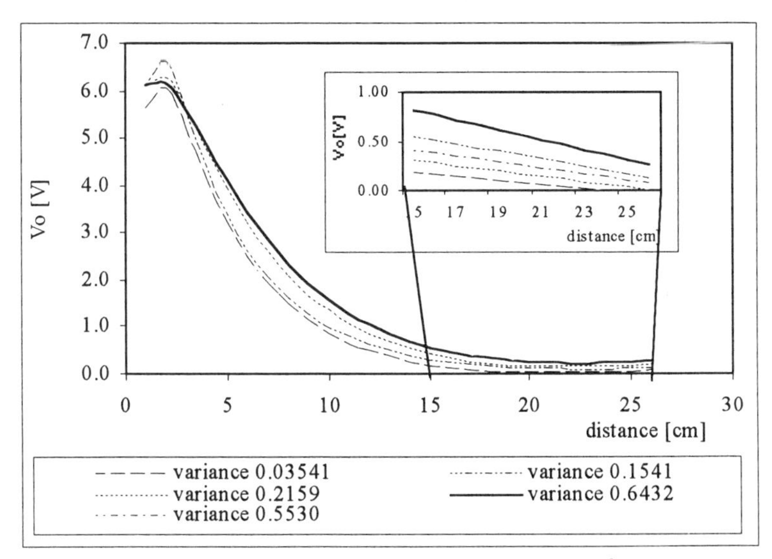

Figure 36. Calibration curves at different values of the noise variance [V^2] when a *uniform* noise signal, generated by a *CNN*, is used.

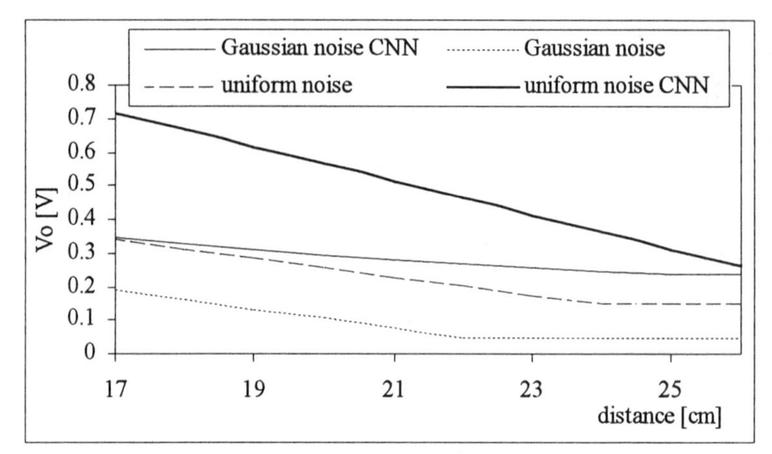

Figure 37. A comparison between the calibration curves obtained in the previous cases.

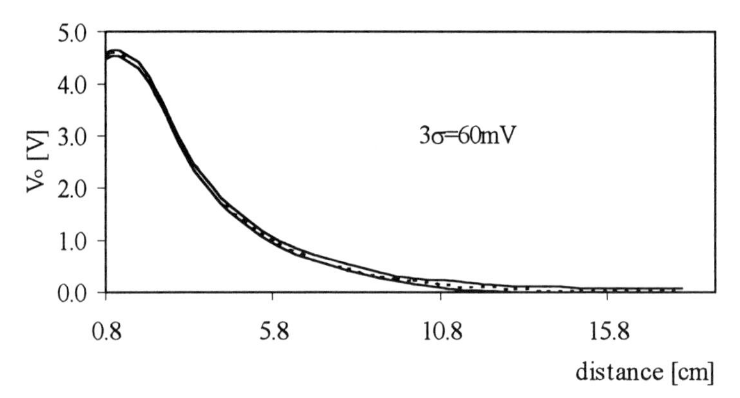

Figure 38. Calibration diagram of the *IR* sensor, without noise modulation. 3σ limits is used.

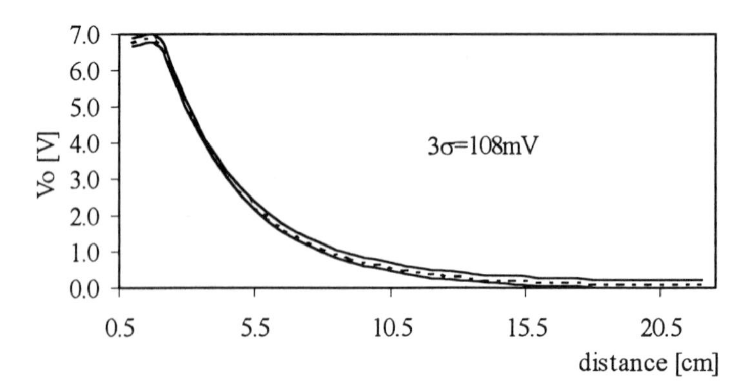

Figure 39a. Calibration diagram when a *Gaussian* noise signal, by a digital generator, is used.

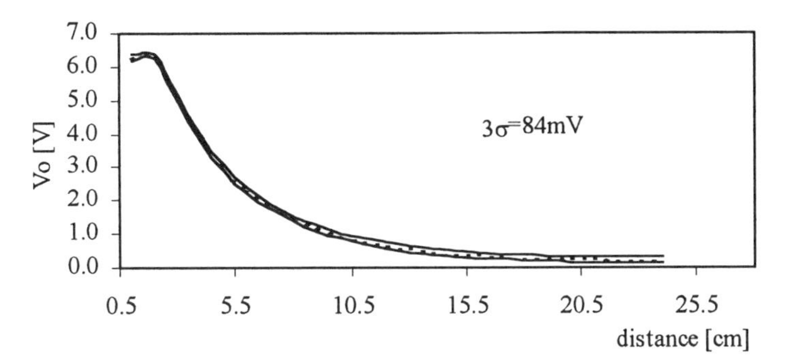

Figure 39b. Calibration diagram when a *uniform* noise signal, by a digital generator, is used.

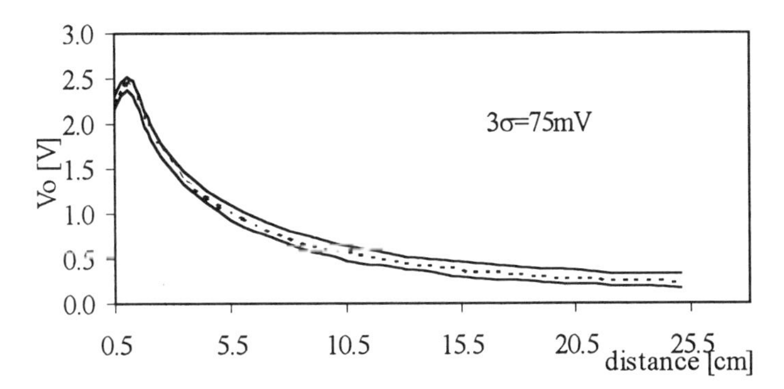

Figure 40a. Calibration diagram when a *Gaussian* noise signal, by a *CNN*, is used.

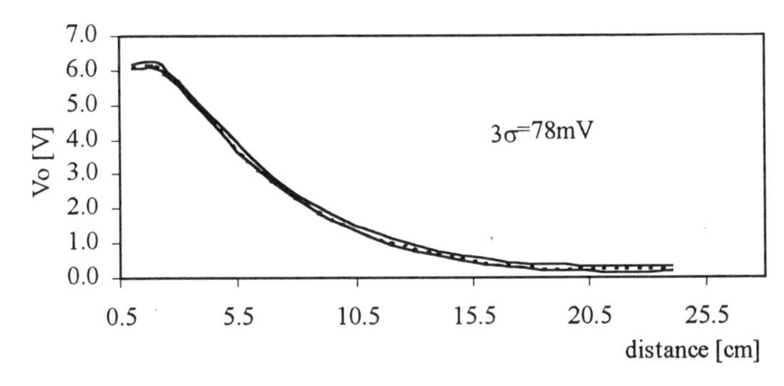

Figure 40b. Calibration diagram when a *uniform* noise signal, by a *CNN*, is used.

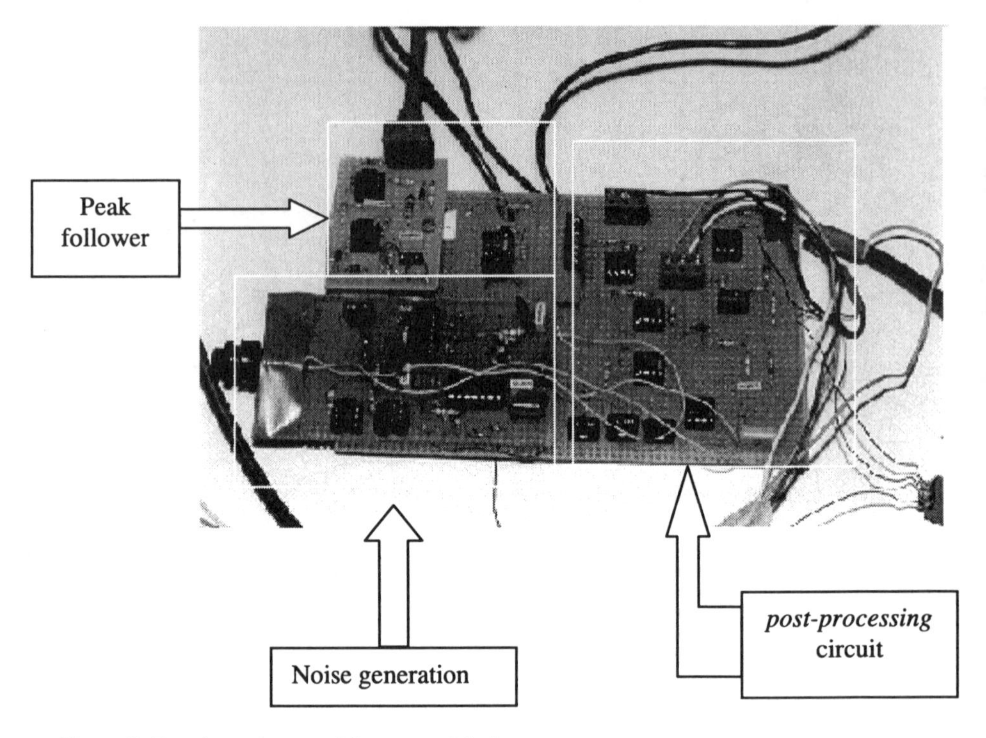

Figure 41. Experimental set-up of the proposed device.

5.7 Dithering in an A/D converter

The example given in this section is to ascertain whether the application of dithering to an A/D converter improves its performance in terms of resolution or an increase in the number of effective bits [10-11]. The system used for the experiment is shown in Figure 42.
The signal passes through the anti.-aliasing filter and reaches the node where it is added to the dither signal and then sampled by means of a sample-and-hold at a sampling frequency, $T_{S/H}$, depending on the ADC conversion time.
The ADC we used is an 8-bit successive approximation converter (*SAR*) whose 600 kHz clock was implemented by means of an external RC circuit. The conversion time, *Tc*, is variable and in the worst case is equal to $T_c=80*T_{CLK}\approx 130\ \mu s$.
Synchronism between the S/H and the ADC is obtained by using a monostable circuit with NAND gates. The pulse supplied by the ADC lasts about 300ns, which is insufficient for the S/H to take samples; the monostable circuit, on the other hand, generates a pulse, *Vm*, of a finite duration $T_a\approx 10\ \mu s$, which is capable of driving the S/H phase correctly. The S/H output signal is also used to enable

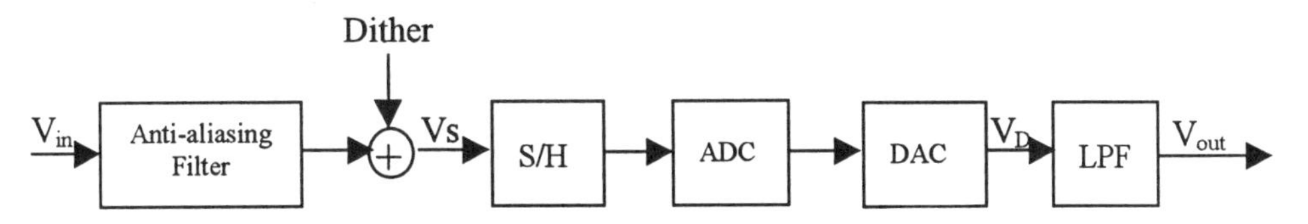

Figure 42. The adopted experimental set-up.

A/D conversion during the hold phase.
For the dither signal, *uniform* noise was used with an amplitude equal to the quantisation step $Q=V_{FS}/2^n$ = 5/256 =19.53 mV, where V_{FS} is the maximum value allowed to the input signal.
To characterise the system the ADC output was reconverted into an analog signal by means of an 8-bit DAC. As illustrated in Chapter 2, the system output was suitably filtered. The system response to a sinusoidal input signal with an amplitude of 120 mV, a frequency of 100 Hz and an offset of 60 mV (an offset voltage had to be added because the AD only converts signals in the range 0-5V) is shown in Figures 43-45. The behaviour of the converter with a *uniform* dither signal with an amplitude equal to the quantisation step is shown in Figures 46-48, where it is possible to see that the dithered system gives better resolution.

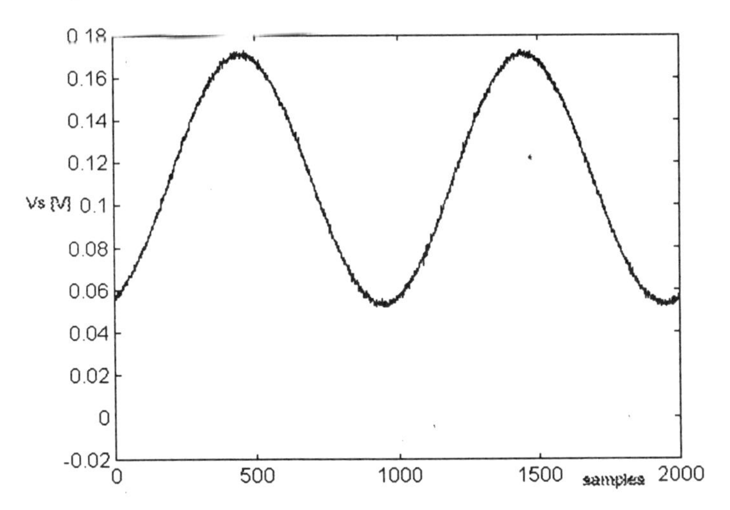

Figure 43. Vs behaviour when no dither signal is applied to the system.

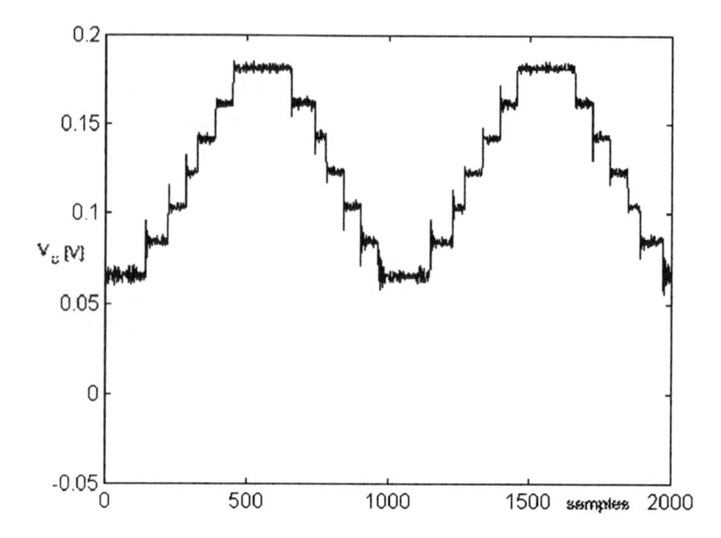

Figure 44. V_D behaviour when no dither signal is applied to the system.

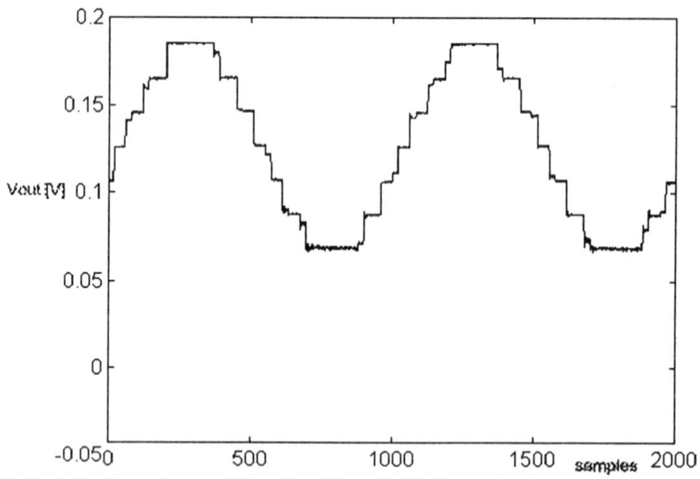

Figure 45. V_{out} behaviour when no dither signal is applied to the system.

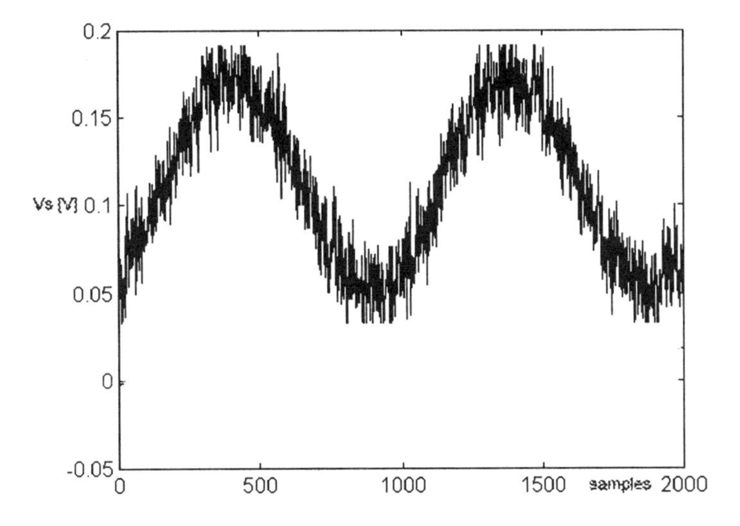

Figure 46. Vs behaviour when a suitable *uniform* dither signal is applied to the system.

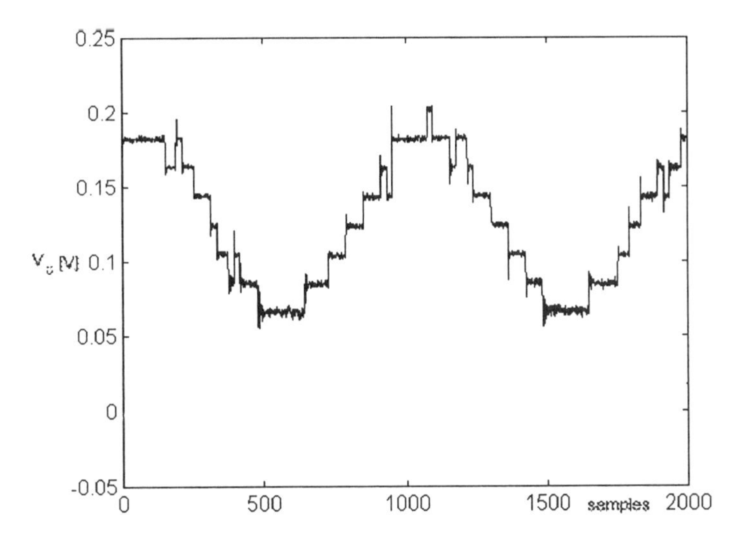

Figure 47. V_D behaviour when a suitable *uniform* dither signal is applied to the system.

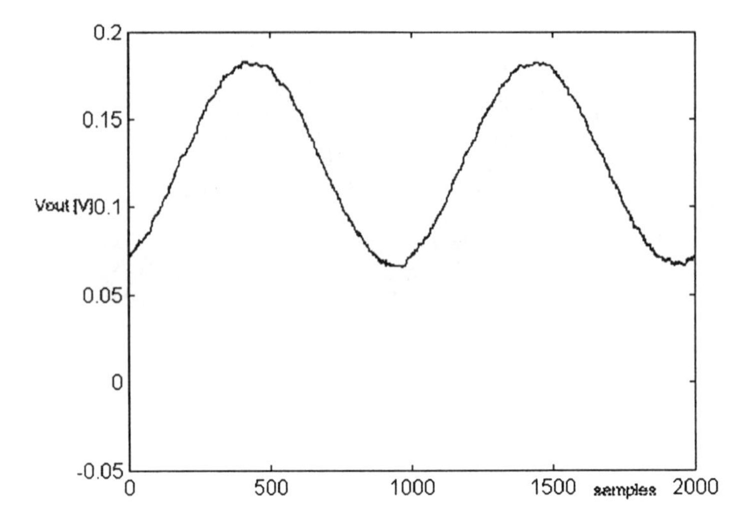

Figure 48. V_{out} behaviour when a suitable *uniform* dither signal is applied to the system.

5.8 Linearisation of an Hall effect switching sensor by noise modulation

The case dealt with in this section is a further example of the use of added noise to modify the characteristic of an essential non-linear device. More specifically, using a Hall effect sensor supplying ON/OFF information, according to the presence or absence of a magnetic flux density higher than 5 mT, we will demonstrate how it is possible to transform the sensor into a linear device for distance measurement.
Using a magnet that with a flux density of 5 mT, we observed the sensor characteristic shown in Figure 49a. As can be seen, the sensor's range of actions about 3 mm.
The aim is to transform the sensor characteristic into one similar to that shown in Figure 49b, thus transforming the device into a proximity sensor. To subject the system to a variable magnetic field (specifically by adding a *uniform* noise with a suitable amplitude) the magnet was replaced by a cylindrical nucleus of soft iron (with a high degree of magnetic permeability) wrapped in a large number (over 1000) of coils. By circulating electrical current in the coil a magnetic flux, B, is generated which can be used to force the sensor. To characterise the sensor, the coil was fixed with a micrometric screw with which it was possible to vary the distance between the sensor and the magnetic source.

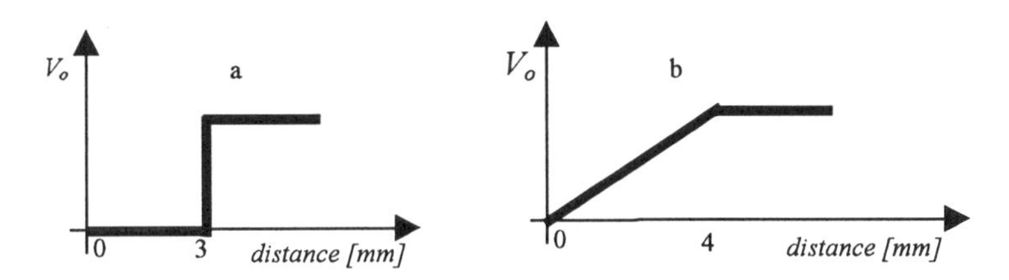

Figure 49. (a) The original sensor characteristic. (b) The sensor characteristic when a suitable noise modulation process is used.

The waveform sent to the coil is an random signal with a *uniform* distribution, whose amplitude and frequency were optimised to maximise the linearity of the characteristic. In these conditions switching is observed at the sensor output, the statistical frequency of which increases as the distance between the nucleus and the sensor decreases; with a post-processing section like the one illustrated in Chapter 2, from this information it is possible to calculate the distance between the sensor and the nucleus. The conditioning circuit of the sensor is shown in Figure 50.
A calibration diagram of the device in noise-added conditions is given in Figure 51. As can be seen, the characteristic is linearised in the range 0-4 mm, with an 3σ limits imprecision of 0.06 mm.

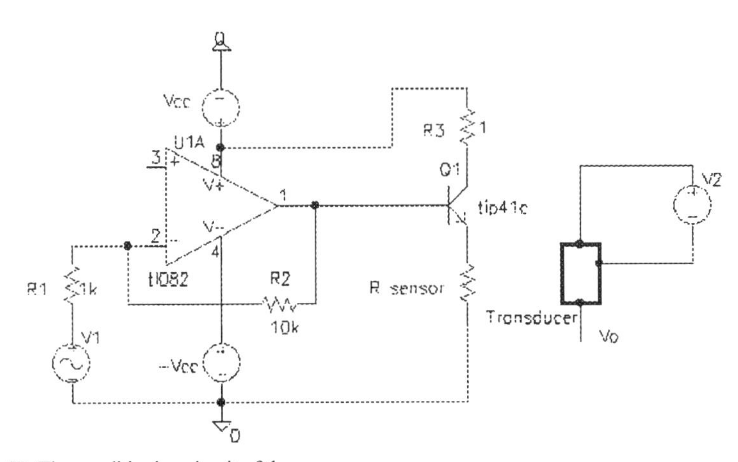

Figure 50. The conditioning circuit of the sensor.

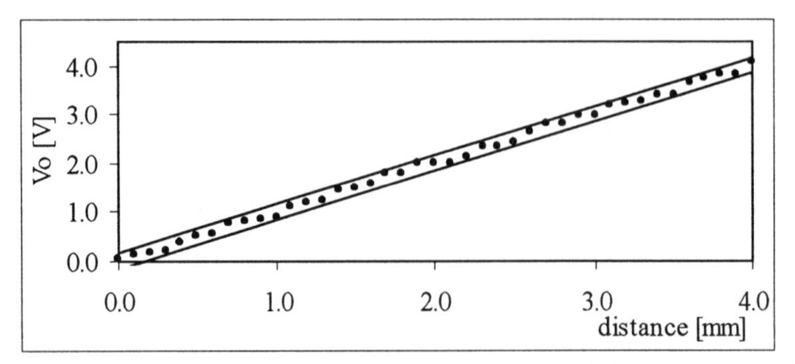

Figure 51. Calibration diagram of the device when a suitable noise modulation technique is used.

5.9 A teaching device for the comprehension of noise acted-systems

In Chapter 2 we introduced the Brownian system as an important example of *SR*. We investigated its most significant properties, its behaviour and the dependence of the latter on the parameters *a*, *b* and γ.
The aim of this section is to describe an analog implementation of the *QDW* so as to have an electronic system analogous to the Brownian one that will illustrate the system's dependence on its characteristic parameters and, at the same time, will allow us to study it not only through numerical simulation but also on the basis of experimental evidence.

5.9.1. ANALOG IMPLEMENTATION OF THE *QDW*

The analog implementation of the system is based on the circuit scheme shown in Figure 52. From the equations at the circuit nodes we get:

$$(u - Vc_1)/R_1 = C_1\dot{V}c_1 + Vc_1/R_3 + (Vc_1 + Vc_2)/R_2 \tag{5.9}$$

for the first node

where: $u = f + Vc_2^3$

and

$$Vc_1 / R_3 = -C_2\dot{V}c_2$$

for the second node.
Rewriting these equations we get:

$$\begin{aligned} \dot{V}c_1 &= -Vc_1(1/R_1C_1 + 1/R_2C_1 + 1/R_3C_1) - Vc_2/R_2C_1 + u/R_1C_1 \\ R_3C_2\dot{V}c_2 &= -Vc_1 \end{aligned} \tag{5.10}$$

Comparing equations (5.10) with (2.21) we get the following correspondence between the circuit parameters and the system parameters

$$
\begin{aligned}
k^2 &= 1/\,R_3C_2 \\
\gamma k &= 1/R_1C_1 + 1/R_2C_1 + 1/R_3C_1 \\
a &= 1/\,R_2C_1 \\
b &= 1/\,R_1C_1
\end{aligned}
\qquad (5.11)
$$

Having assigned the system parameters a set of values commonly used in the literature

$$
\begin{aligned}
\gamma &= 0.236 \\
a &= b = 0.967
\end{aligned}
$$

we get

$$
\begin{aligned}
R_1 &= R_2 = 470k\Omega \\
R_3 &= 1k\Omega \\
C_1 &- 2.2\mu F \\
C_2 &= 270pF \\
(Rc &= 470k\Omega)
\end{aligned}
$$

This analog system was first simulated in a *SPICE®* environment and then analogically implemented. Figures 53-58 illustrate the evolution in time and state diagrams of the system, obtained via simulation with different periodic forcing signal values. Likewise, Figures 59-64 illustrate the experimentally observed evolution of the state variables and the state diagrams of the analog system.
Forcing the system with a signal comprising a periodic component and a *Gaussian* noise it was possible to observe the behaviour discussed in detail in Chapter 2.

5.9.2 A DEVELOPMENT ENVIRONMENT FOR THE STUDY OF PHENOMENA CONNECTED WITH THE *QDW*

To assess the effects of parametric variations on the behaviour of the system, it is useful to adopt a strategy that will allow us to intervene interactively on the parameters of interest and observe the system's response. The parameters taken into consideration are those modulating the shape of the potential *V(x)* and the product γK of the friction coefficient and the scaling factor.
To vary these parameters we used three potentiometers which, acting on the values of the resistance coils R_1, R_2 and R_3, allow us to model the shape of the

potential, i.e. the distance between the two wells and the height of the barrier. From Equation (5.11) it can be seen that parameter a depends on R_2, parameter b on R_1 and k on R_3. Finally, it can be hypothesised (with suitable simplifications) that parameter γk depends exclusively on R_3. Due, in fact, to the values taken by the resistance coils R_1, R_2 and R_3, we can write

$$\gamma k = 1/R_1C_1 + 1/R_2C_1 + 1/R_3C_1 \approx 1/R_3C_1$$

and therefore only act on R_3 to vary the product.
With the ranges of variation chosen for the resistance value, we can act on the system parameters as follows

$a=0.31\div1.39$
$b=0.68\div4.59$
$k=628\div8784$
$\gamma k=48.4\div9469$

Therefore, by acting on the organs controlling the parameters a, b and γ, it is possible to change their characteristics in such a way as to observe the behaviour discussed in Chapter 2.
To visualise the effect of parametric variations on the shape of the potential $V(x)$ a virtual tool was implemented using the *LabVIEW®* code, which gives the user a clear, efficient representation of the state of the system. The front panel of the tool is shown in Figure 65. As can be seen, the tool not only shows the evolution of the potential $V(x)$ but also gives information about the values of the system parameters. The interface between the virtual tool and the analog device implementing the *QDW* is a data acquisition board (DAQ). The logical scheme on which the tool is based is shown in Figure 66.
One of the investigations performed to test the effectiveness of the structure, measured the values of the amplitude A and the frequency ν of the forcing signal needed to make the system switch, varying the magnitudes characteristic of the potential DV and $x_\pm$.
The result of the analysis is summarised in Figures 67 and 68. The graphs give the optimal values for the distance between the two wells, x_m, in two conditions: a constant amplitude with a varying frequency and a constant frequency with a varying amplitude. As observed via simulation in Chapter 2, to guarantee optimal system performance it is necessary, as the frequency increases, to reduce the distance between the two wells; likewise, as the amplitude A increases, the distance has to be increased.
Besides its usefulness as a technique for experimental investigation, the system is of great interest for teaching purposes. Observing on a screen of an oscilloscope how the state trajectories of a complex system like the *QDW* are modified when the shape of the potential changes (this can also be observed through a user

interface) is of significant help in understanding such complex phenomena. Besides the fundamental support of the theoretical background and confirmation received via simulation, in fact, students or researchers studying complex phenomena frequently require confirmation via experimentation. In this sense the synergy between practice and theory, hardware and software, an analog implementation and a synoptic interface, provide a concrete completion of the process of cognitive acquisition of certain classes of phenomena. In this context, tools like the one discussed here play a fundamental role in the understanding of added-noise systems and their sensitivity to parametric variations.

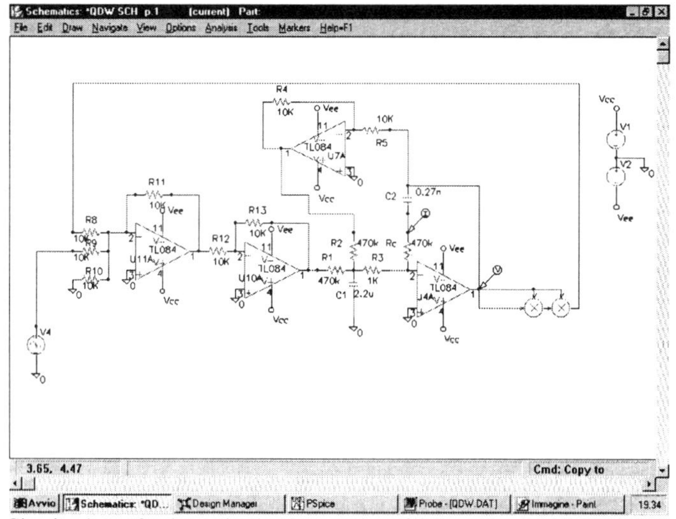

Figure 52. Circuit scheme for the analog implementation of the *QDW* system.

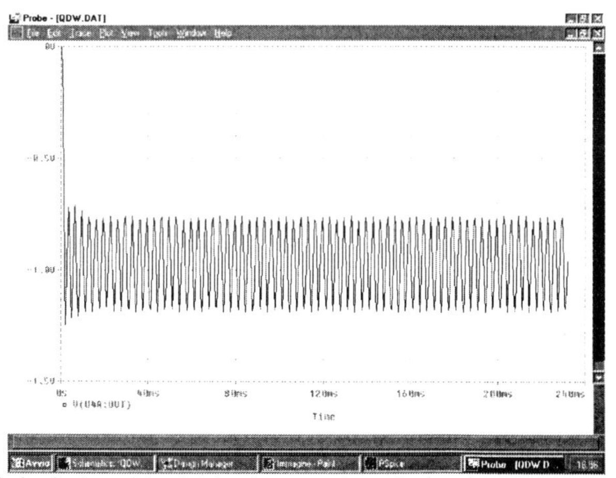

Figure 53. Evolution in time of the simulated system when a forcing signal with A=200mV is used.

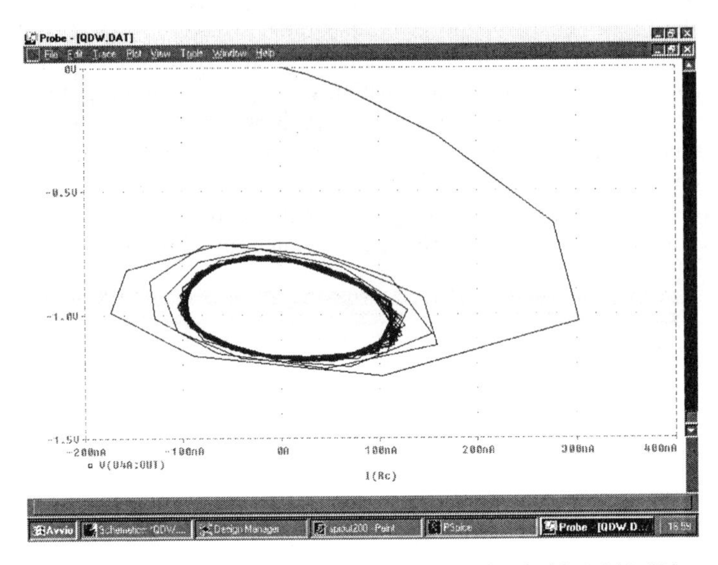

Figure 54. State diagram of the simulated system when a forcing signal with A=200mV is used.

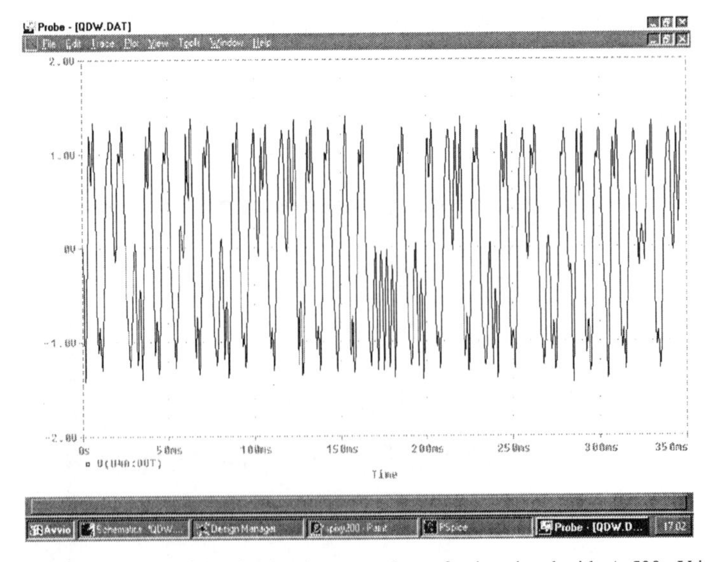

Figure 55. Evolution in time of the simulated system when a forcing signal with A=500mV is used.

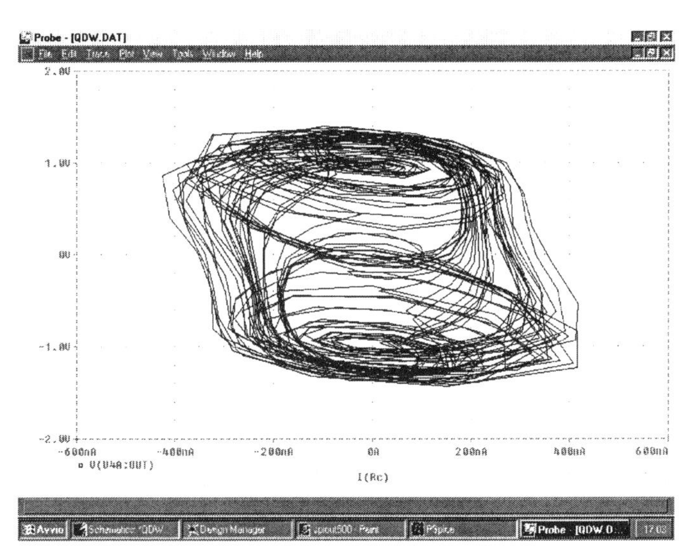

Figure 56. State diagram of the simulated system when a forcing signal with A=500mV is used.

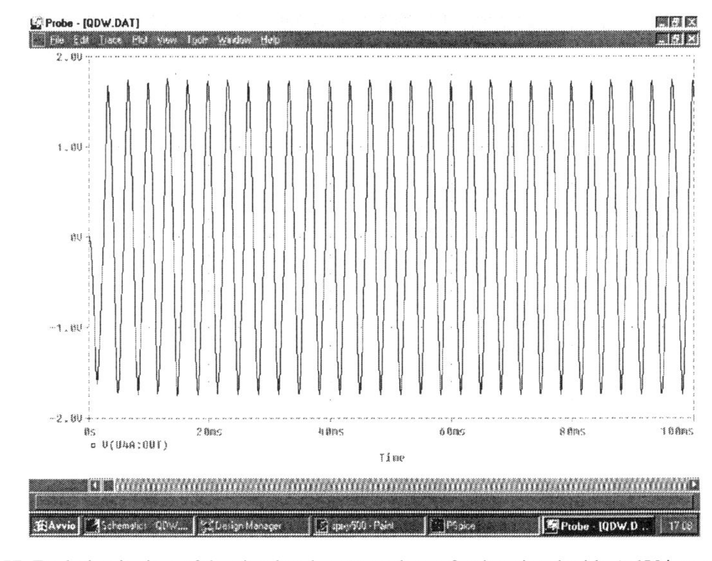

Figure 57. Evolution in time of the simulated system when a forcing signal with A=1V is used.

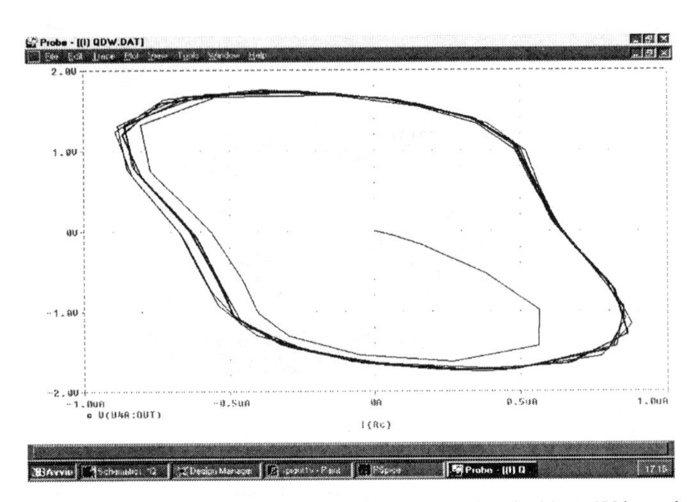

Figure 58. State diagram of the simulated system when a forcing signal with A=1V is used.

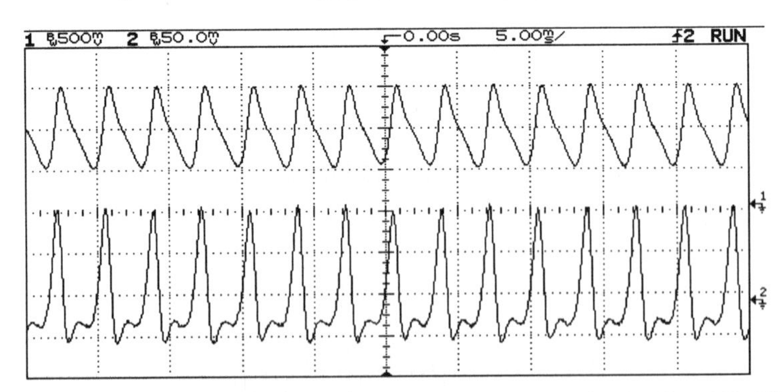

Figure 59. Evolution in time of the experimental system when a forcing signal with A=550mV is used.

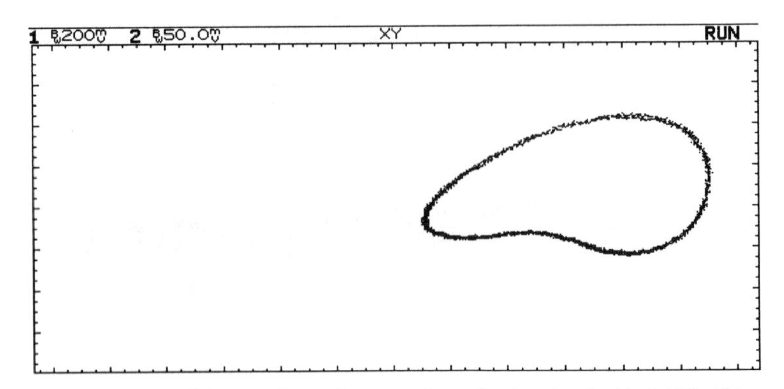

Figure 60. State diagram of the experimental system when a forcing signal with A=550mV is used.

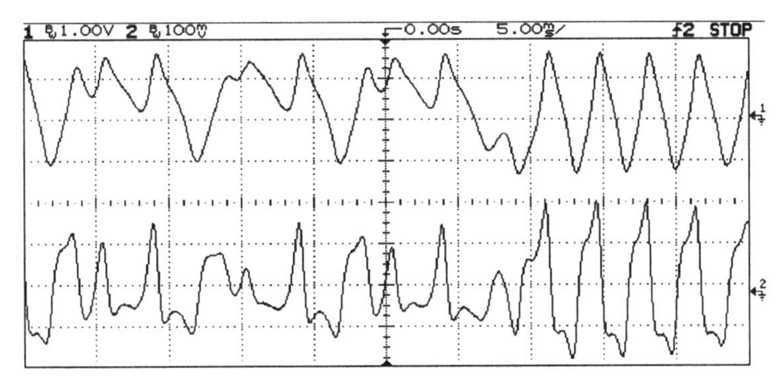

Figure 61. Evolution in time of the experimental system when a forcing signal with A=600mV is used.

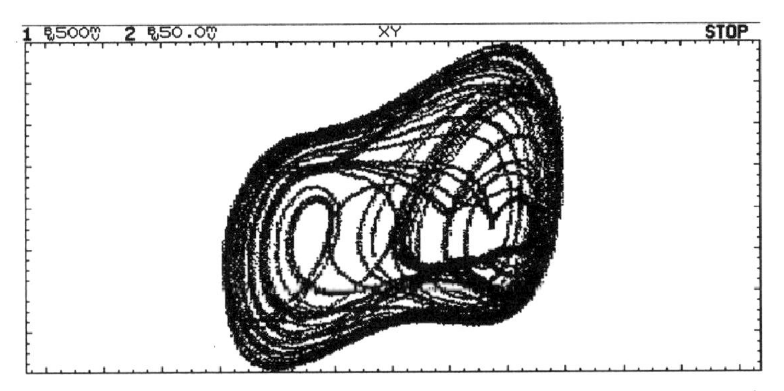

Figure 62. State diagram of the expeerimental system when a forcing signal with A=600mV is used.

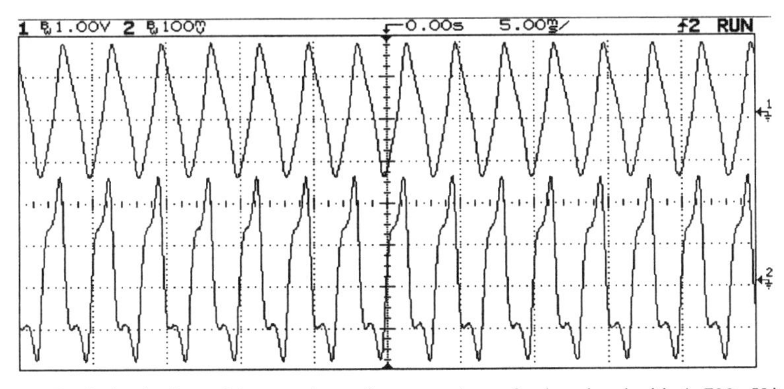

Figure 63. Evolution in time of the experimental system when a forcing signal with A=700mV is used.

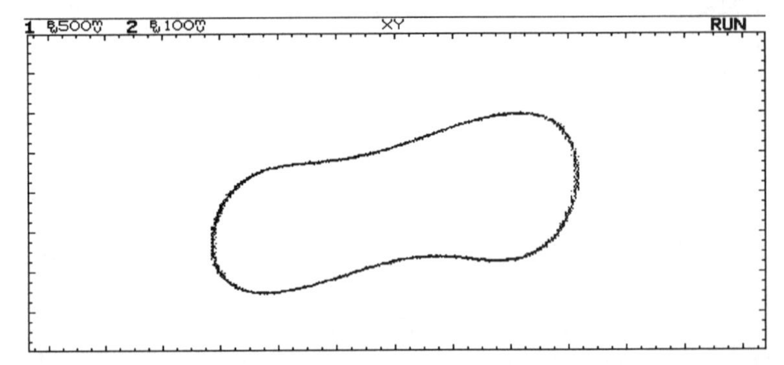

Figure 64. State diagram of the simulated system when a forcing signal with A=700mV is used.

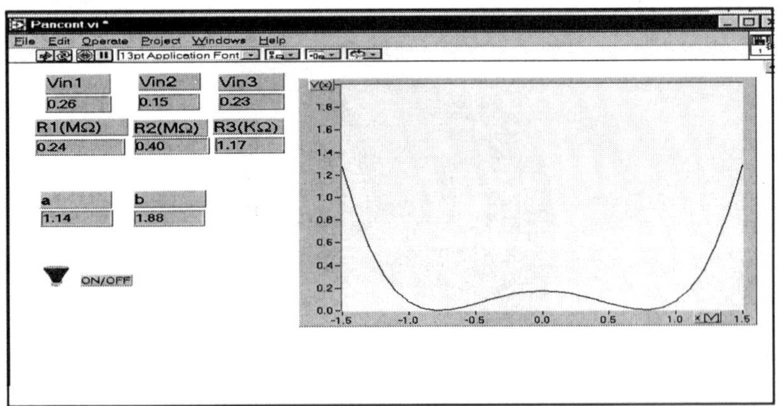

Figure 65. Front panel of the virtual instrument allowing for *QDW* system analisys.

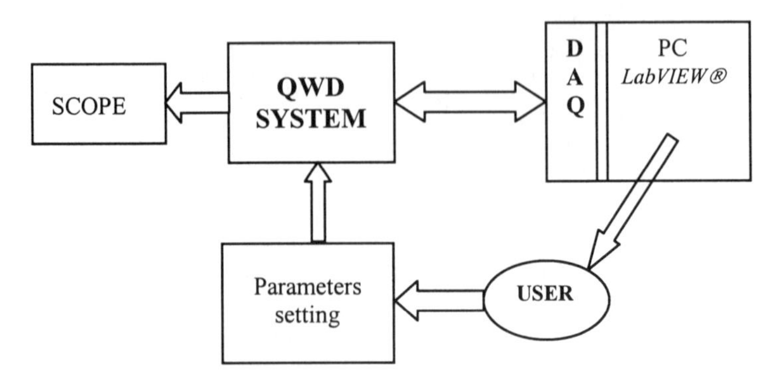

Figure 66. Logical scheme of the teaching tool.

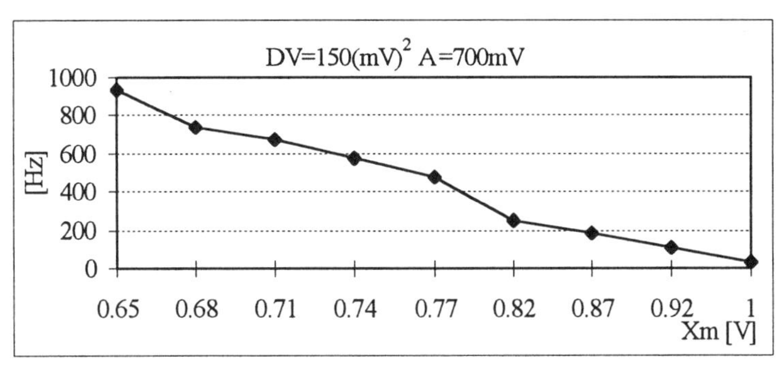

Figure 67. Interwell distance at different values of the forcing signal frequency.

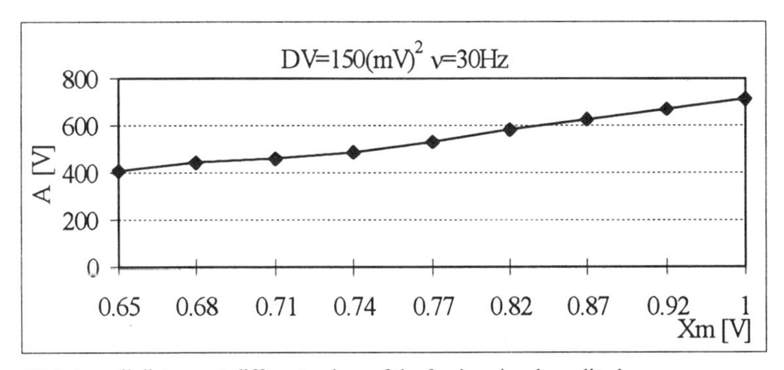

Figure 68. Interwell distance at different values of the forcing signal amplitude.

5.10 Signal recovery based on a *QDW* approach

We will now present an application based on a QDW system which reveals periodic tones completely hidden by a stochastic component. The latter is assumed to have an exclusively *Gaussian* distribution and no hypothesis is made concerning the value of the statistical moments.
The main idea investigated here is the possibility of moving a system into the *SR* condition when it is forced by an unknown periodic signal modulated by a white *Gaussian* noise with unknown statistical properties, by changing its structural parameters a and b. It should be noted that in this case the damping factor γ should be kept at a low value in order to run the underdamped mechanism in the system. When the *SR* condition is reached the required information on the periodic forcing frequency can be obtained.
If the *QDW* system is forced by a periodic signal and a noise, the waiting time between two transactions can be monitored. The distribution of this quantity shows several peaks decreasing in amplitude. The first peak corresponds to the forcing frequency 2ν while the other peaks are odd multiples of ν. The area E

under the first peak corresponds to the probability that the system is synchronised with the periodic signal. Of course the value of E changes according to the system parameter. For example if the noise variance is considered, E will reach a maximum value for $\sigma=\sigma_{opt.}$ Similar considerations can be made on the possibility of synchronising the system by acting on the parameters a and b, the other quantities being fixed. Hence for a suitable range of these parameters the waiting time distribution will show a maximum corresponding to the forcing frequency and the area E under this maximum will be magnified with optimal values of a and b.

The residence time distribution is strictly related to the power spectrum of the system output (i.e. the state x). It is interesting to observe that if the quantity E is plotted as a function of the forcing frequency ν it shows typical resonant behaviour, confirming that when all the stochastic system parameters (noise variance, forcing amplitude, structural parameters) are fixed a resonant frequency can be defined [19].

On the basis of the above considerations the following rules can be obtained:

- when all the stochastic system parameters are suitably fixed, the power spectrum of the system output will show a peak corresponding to the forcing frequency;
- this peak will reach a maximum when the system parameter values are optimal.

The fact that the *QDW* system is extremely sensitive to the forcing harmonic, even in presence of noise, is the main factor in the proposed recovery methodology.

The procedure starts with the computation of the system output for a large set of values of a and b. The algorithm adopted computes the frequency corresponding to the peak value for each spectrum, computes the distribution of these frequencies and chooses the frequency with the maximum number of occurrences. It has been experimentally proved that this maximum value is reached at the forcing frequency. The core of the developed system is the *Matlab®* environment.

5.10.1 A VIRTUAL INSTRUMENT FOR NOISE CORRUPTED HARMONIC RECOVERY

A virtual instrument allowing the user to be confident with the selection algorithm has been developed in *LabVIEW®*. The instrument uses the *DDE* library in order to establish a communication with the *Matlab®* code. The front panel of the instrument in different study cases is shown in Figures 69-74. The instrument is divided into two sections. The top section is dedicated to the input/output data actions while the bottom section is devoted to the graphic output.

In the data input section the user has to insert the parameters of the periodic forcing signal (amplitude and frequency) and the stochastic signal. At this point,

having selected the selection box on the upper left-hand side the signal is processed.
In the middle section the spectra of the *QDW* system output for a large set of structural parameters are given.
To check the utility of the selection algorithm a classical analysis of the input signal was made. In the top section of the graphic area the power spectrum, the correlation map and the statistical distribution of the signal are shown. It should be observed that by using the approaches mentioned no information on the forcing signal frequency can be obtained.
The next step is to apply the algorithm to determine the unknown pulse υ_x. To do so, it is necessary to act on the selection box on the top right-hand side of the front panel.
In the bottom section the spectra of the output showing the maxima in a narrow range of the frequencies selected by the algorithm are presented.
The detected frequency is shown in the top right-hand dialog box and in the test mode it can be compared with the input frequency given in the data section, in order to verify the efficiency of the tool.

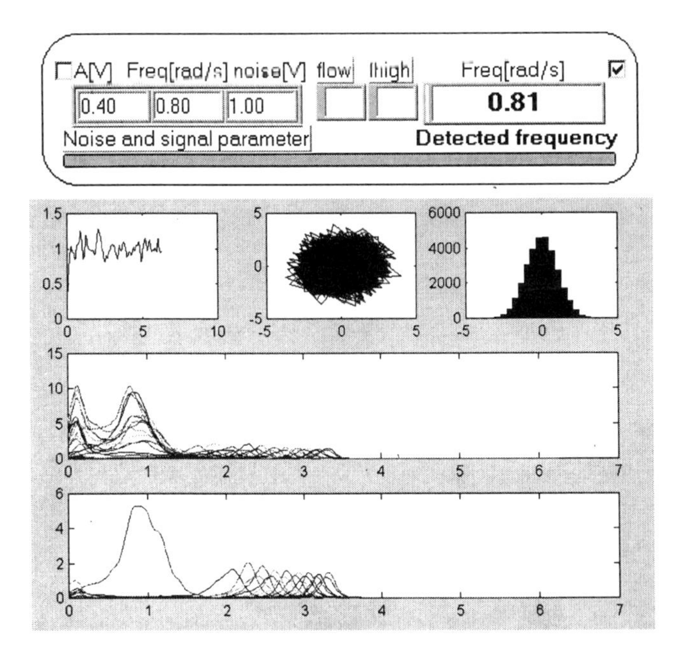

Figure 69. Front panel of the virtual instrument for harmonic recovery, in case of υ_x=0.8 rad/s. The instrument uses the *DDE* library in order to establish a communication with the *Matlab*® code. The instrument is divided into two sections. The top section is dedicated to the input/output data actions while the bottom section is devoted to the graphic output.

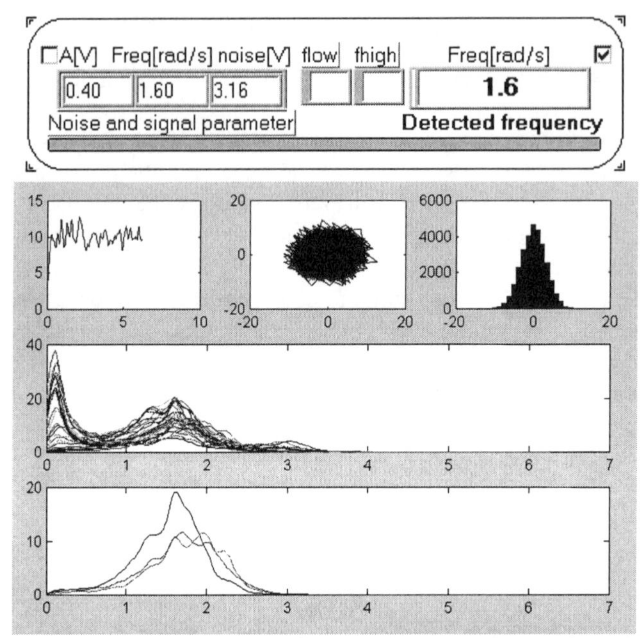

Figure 70. Front panel of the virtual instrument for harmonic recovery, in case of $\upsilon_x = 1.6$ rad/s.

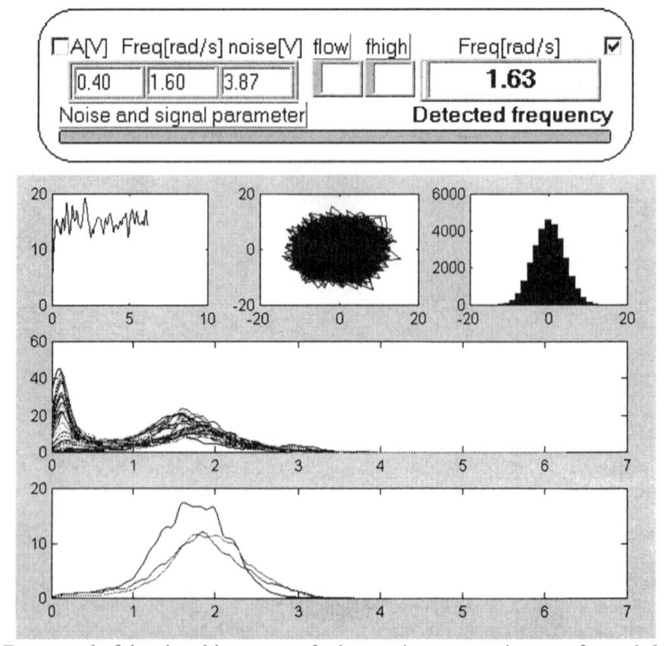

Figure 71. Front panel of the virtual instrument for harmonic recovery, in case of $\upsilon_x = 1.6$ rad/s.

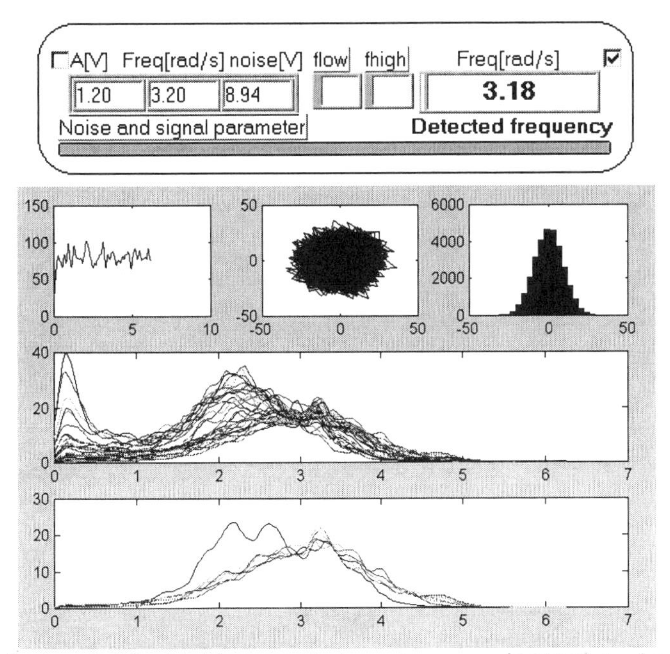

Figure 72. Front panel of the virtual instrument for harmonic recovery, in case of $\upsilon_x = 3.2$ rad/s.

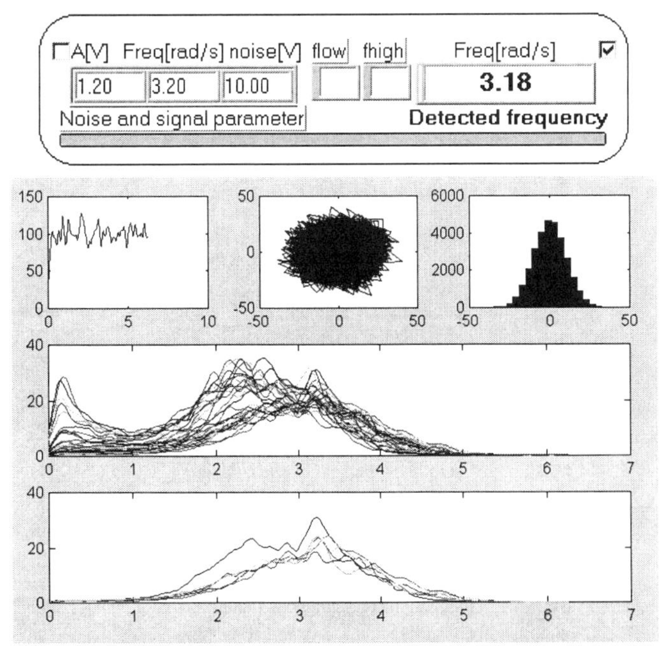

Figure 73. Front panel of the virtual instrument for harmonic recovery, in case of $\upsilon_x = 3.2$ rad/s.

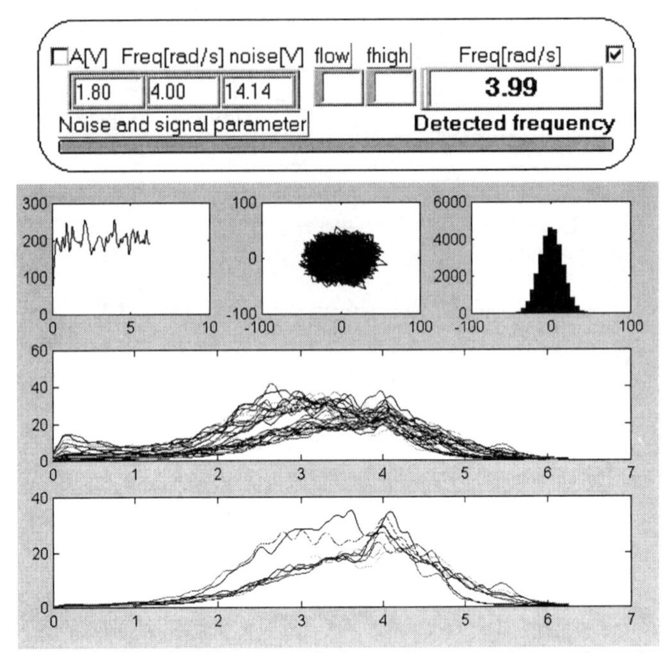

Figure 74. Front panel of the virtual instrument for harmonic recovery, in case of υ_x = 4 rad/s.

References

[1] L. Gammaitoni, F. Marchesoni, E. Menichella-Saetta, and S. Santucci, "Stochastic Resonance in Bistable Systems", *Phys. Rev. Lett. 62, 1989, 49.*

[2] Roberto Benzi, Alfonso Sutera and Angelo Vulpiani, "The mechanism of stochastic resonance", *J. Phys. A: Math. Gentile. 14, 1981, L453.*

[3] S. FAUVE and F. Heslot, "Stochastic Resonance in a Bistable Systems", *Phys. Lett. 97A,1983, 5.*

[4] B.Andò, S. Baglio, S. Graziani, N. Pitrone, "A probabilistic approach to the threshold error reduction theory in bistable measurement devices", *IMTC98, S. Paul, Minnesota, 1998.*

[5] B. Andò, S. Baglio, S. Graziani, N. Pitrone, "Characterisation of threshold error via stochastic resonance", in: IMEKO '97, Helsinki, 1997

[6] B. Andò, S. Baglio, S. Graziani, N. Pitrone, "Virtual instruments with low threshold error based on stochastic resonance theory", *SICICA '97, Annecy, France, 1997.*

[7] B.Andò, S. Baglio, S. Graziani, N. Pitrone., 1999, "Optimal improvement in bistable measurement device perfromance via stochastic resonance" INT. J. ELECTRONICS, vol 86, n. 7.

[8] B. Andò, S. Baglio, S. Graziani, N. Pitrone, a system for the implementation of noise added System driving, IMTC'99, Venezia, 1999.

[9] B.Andò, S. Baglio, S. Graziani, N. Pitrone, " Threshold error reduction in linear measurement devices by adding noise signal", *IMTC98, S. Paul, Minnesota, 1998.*

[10] Carbone P. and Petri, D., 1994, "Effect of Additive Dither on the Resolution of Ideal" Quantizers. *IEEE Transactions on instrumentation and measurement* **43**.

[11] Carbone, P., Narduzzi, C., and Petri, D., 1996, "Performance of Stochastic Quantizer Employing Nonlinear Processing." *IEEE Transactions on Instrumentation and Measurement* **45**.

[12] Gammaitoni, L., 1995, "Stochastic Resonance and the dithering effect in threshold physical systems." *Phys. Rev. E* **52**.

[13] Gammaitoni, L., Hanggi, P., Jung, P., and Marchesoni, F., 1998, "Stochastic Resonance." *Rev. of Modern Physics* **70**.

[14] Ernest O. Doebelin, "Measurement Systems", *McGRAW-HILL BOOK COMPANY*, third edition, 1985.

[15] L.O. Chua, M. Hasler, G.S. Moschytz, J. Neirynck, "Autonomous Cellular, "Neural Networks: A Unified Paradigm for Pattern Formation and Active Wave Propagation", IEEE Trans. on Circuits and Systems - Part I, vol. 42, no.10, 1995, pp.559--577.

[16] P. Arena, S. Baglio, L. Fortuna, G. Manganaro, "Self Organization in a two-layer CNN", IEEE Trans. on Circuits and Systems - Part I, vol. 45, no.2, Feb.1998, pp.157-162.

[17] P. Arena, Riccardo Caponetto, Luigi Fortuna, Alessandro Rizzo, "Noise supported Wavefronts in Cellular Neural Network Based Circuits", (to be printed) IEEE Circuits and System.

[18] Gammaitoni L., Marchesoni F., Santucci S., "Stochastic Resonanc as a Bona Fide Resonance", *Phys. Rev. 74/7, 1995, 1052.*

[19] José Brandao Faria, "A theoretical Analysis of the Bifurcated Fiber Bundle Displacement Sensor", IEEE Trans. On Instr. And Measur., Vol. 47, 1998.

6 THE NASS SIMULATION ENVIRONMENT

6.1 Introduction

The analysis of added-noise techniques, as presented in the previous chapters, has clearly demonstrated the complexity of the topics being dealt with. More specifically, the different modes of behaviour of linear and bistable systems, or again the complexity of techniques to determine optimal noise amplitudes, suggest the use of simulation environments that make it possible to perform a qualitative analysis of the behaviour of a stochastic system in a way that is not excessively complex.

A possible development environment for added-noise systems, the *NASS* (Noise-Activated Systems Simulation) code, which the authors developed on the basis of teaching requirements, will be illustrated in this section

6.2 The *NASS* code

The idea on which the software is based is that of implementing a tool that can provide the user with a flexible, interactive simulation environment. The development environment used is the *Matlab®* code.

The tool has two separate sections that can be selected from the initial menu, one referring to linear systems and the other to bistable systems. The latter part is much more complex due to the presence of sections devoted to implementation of the law controlling the noise variance.

The default setting simulates a threshold system with hysteresis as an example of threshold reduction, and a device with pseudo-linear characteristics (thus with an inherent threshold) as an example of linearisation. This is due to the tradition of noise-added techniques, in which such systems are at the basis of the most widespread theories [1, 2]. The initial display is shown in Figure 1. Once the user has selected type of system on which to operate, the user interface changes.

By interacting with the appropriate functions and procedures, however, it is possible to modify the setting and operate with different systems.

In both sections there is an operating mode devoted to manual system parameter setting (threshold amplitude, frequency and amplitude of the forcing signal, noise

intensity, etc.). In the section dedicated to bistable systems there is also an operating mode in which, once all the system parameters except for one have been set, the remaining parameter is set automatically by the optimisation algorithms implementing the law controlling the variance, as discussed in Chapter 3.
In the following sections we will examine the main functions implemented in the *NASS* software for both bistable and linear systems.

6.2.1 BISTABLE SYSTEMS

Figure 2 shows the first display of the *NASS* code section devoted to bistable systems. There are two keys in the selection box: *Control* and *Initialise*.
The initialise Menu allows to show the *minimum pair* and the *iso-amplitude map*, introduced in Chapter 3, for the system being considered. Figure 2 shows the user interface when the *iso-amplitude* function has been selected.
The Control key enables the user to vary the parameters of the stochastic system. The parameter values are automatically updated in the parameter setting box and the system response is displayed in the graphics area.
A typical screen following selection of the Control key is shown in Figure 3. As can be seen, the screen is now divided into three main areas:

- the bottom left-hand side is reserved for user input;
- the upper area is devoted to graphic output, to display the input and output signals of the system and the stochastic forcing signal
- the small box in the center area is reserved for the data output.

The *Frequency*, *Amplitude* and *Noise STD* keys enable the user to set the system parameters.
The graphics area on the right-hand side shows the input signal (the sinusoidal forcing signal with added noise) and the system response (transitions) to these forcing signals.
The *optimise* function gives access to the tuning section of the code which implements techniques for automatic determination of the optimal parameter values. The tuning section is shown in Figure 4.
The *Var and Freq, Ampl and Freq, Ampl and Var* options can be used to determine the remaining parameter (in the first case, for example, the minimum amplitude of the periodic forcing signal) using the control law introduced in Chapter 3.
Once the frequency has been set, the *Optimal System Performance* function makes it possible to force the system with the *minimum pairs* on the basis of a stored archive. For example, in case of a 1KHz forcing frequency the minimum values ($A=0.220V$, $\sigma=0.260V$) are selected. These values are also set in the parameter menu, as can be seen in Figure 5. The graphics area shows how the system switches between states even though it is forced with a signal whose amplitude is lower than the threshold.
Now let us consider Figure 6, in which the amplitude has been altered ($A=0.4V$) as compared with the minimum values. Keeping the forcing signal amplitude and frequency fixed by using relation (3.23) (key *Ampl and Freq*) the code determines the optimal variance value at which the system switches. This situation is

illustrated in Figure 7, where new value for the noise amplitude σ_{opt}=0.220V is suggested by the code. Obviously, as the forcing signal amplitude has been increased above the minimum pair, the optimal standard deviation has decreased. As can be observed in Figure 7, with a value of σ=0.220V the system switches whereas with lower values, σ=0.200V, it cannot follow the dynamic of the periodic forcing signal, as shown in Figure 8.
Similar situations are to be encountered in the case of the standard deviation, σ. Figures 9 and 10 illustrate the case in which the system is forced by a stochastic signal with a standard deviation of σ=0.230V, lower than the optimal standard deviation of σ_{opt}=0.36V.
After execution of the appropriate procedure, the code returns a minimum value for the forcing signal amplitude, i.e. 0.320V. Observing the graphics area in Figure 10, it is possible to confirm the accuracy of the theoretical prediction. Further confirmation is given by observing the response of the system when forced by a signal with an amplitude of 0.300V, i.e. lower than the estimated value, as shown in Figure 11. In this case system transitions are inhibited by the excessively low value of A.

6.2.2 LINEAR SYSTEMS

The initial display of the section devoted to linear systems is shown in Figure 12. As can be seen, the keys in the parameter menu can be used to increase or decrease the system threshold values, the amplitude and frequency of the forcing signal, the amplitude of the stochastic forcing signal and the filter parameters.
The filtering algorithm can be chosen from among a number of well-known algorithms. The stochastic forcing signal chosen was a noise with a *uniform PDF*, a null mean and a standard deviation σ.
Figures 12 and 13 show the evolution of the input and output signals of the linear system, with two different noise amplitudes. The top section of the graphics area shows the characteristics of the original system, i.e. without the additional stochastic source.
The central graph shows the characteristic of the system when subjected to a stochastic forcing signal, whereas the curves in the bottom part refer to the input and output signals.
As can be seen from the IN-OUT characteristics of the system, shown in Figure 12, when the noise amplitude is insufficient the system is unable to process correctly input signals with an amplitude lower than the threshold; vice versa, by suitably modulating the variance of the noise, the system made up of the original system and the filter is a linear one even when the input values are low, as emerges from analysis of Figure 13.
In addition, in perfect accordance with the theories introduced in Chapter 3, the optimal noise standard deviation is $S/\sqrt{3}$.

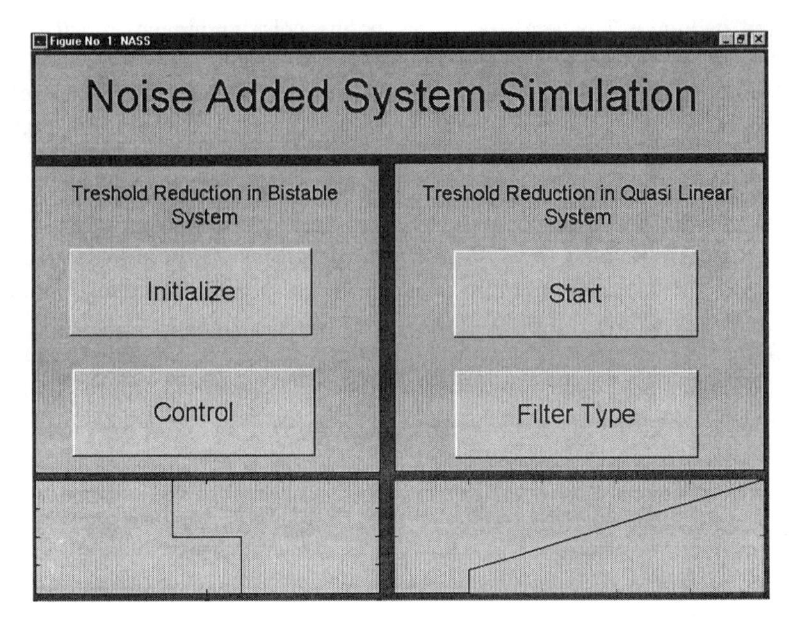

Figure 1. Initial display of the *NASS* code. The default setting simulates a threshold system with hysteresis as an example of threshold reduction, and a device with pseudo-linear characteristics (thus with an inherent threshold) as an example of linearisation.

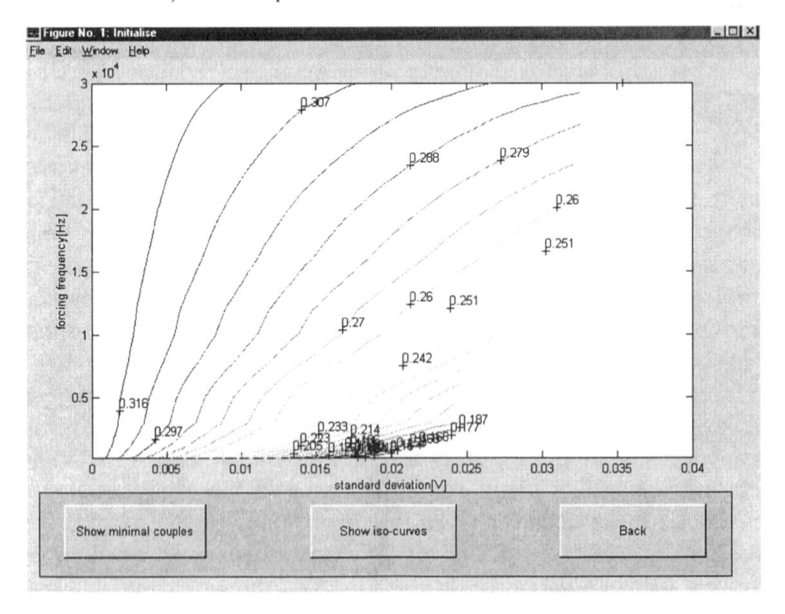

Figure 2. The first display of the NASS code section devoted to bistable systems, when the *iso-amplitude* task has been invocated.

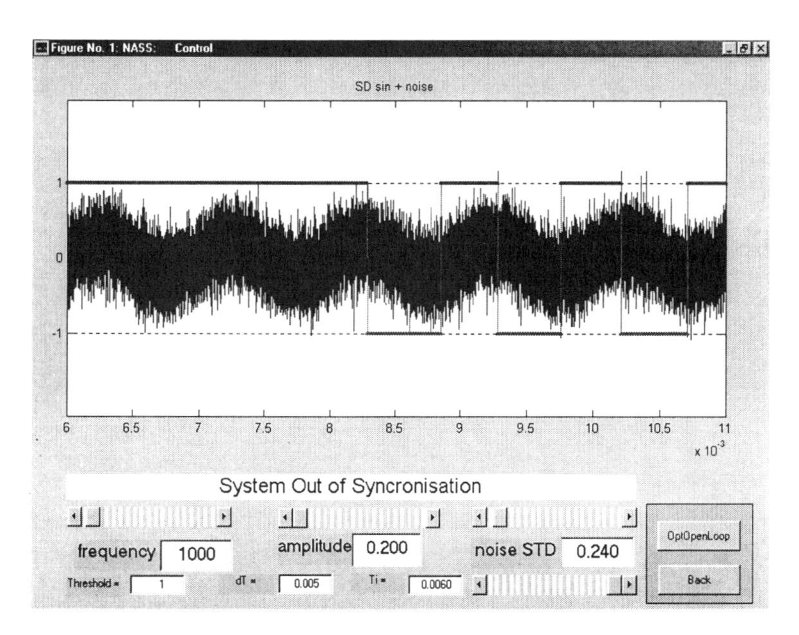

Figure 3. System parameter Control section. The screen is divided into three main areas: the bottom left-hand side is reserved for user input; the upper area is devoted to graphic output, to display the input and output signals of the system and the stochastic forcing signal; the box in the center area is reserved for output data.

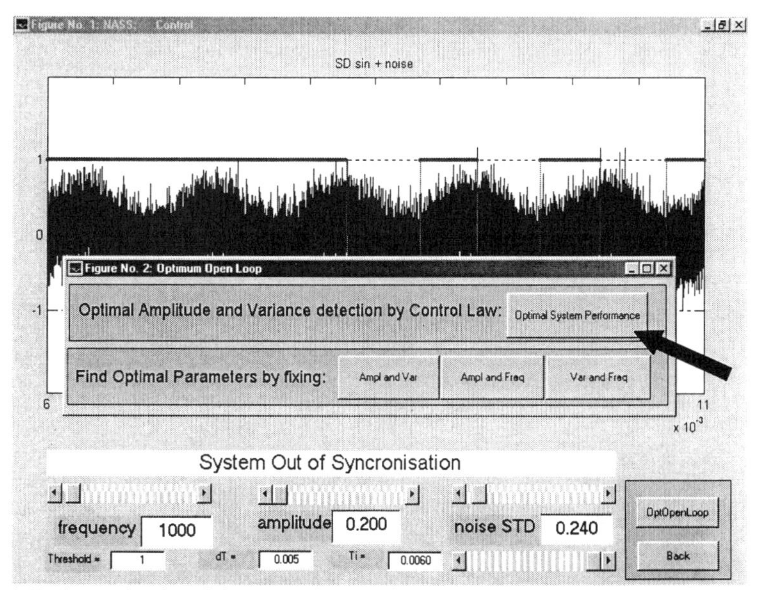

Figure 4. Tuning section for minimum couples detection, invocate by the *OptOpenLoop* function

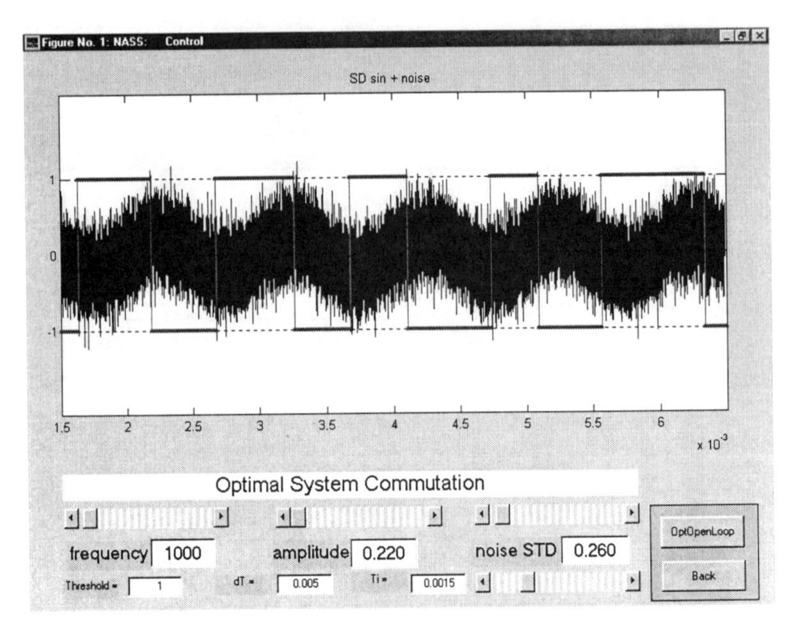

Figure 5. Result obtained by using the *Optimal System Performance* function. The minimum couple has been set.

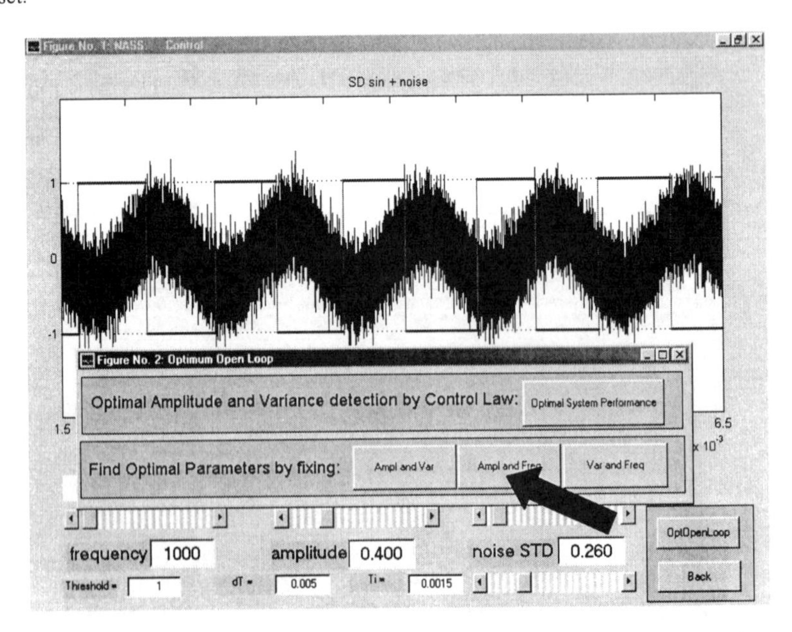

Figure 6. Tuning section for optimal noise variance value detection, when the amplitude has been altered (A=0.4V).

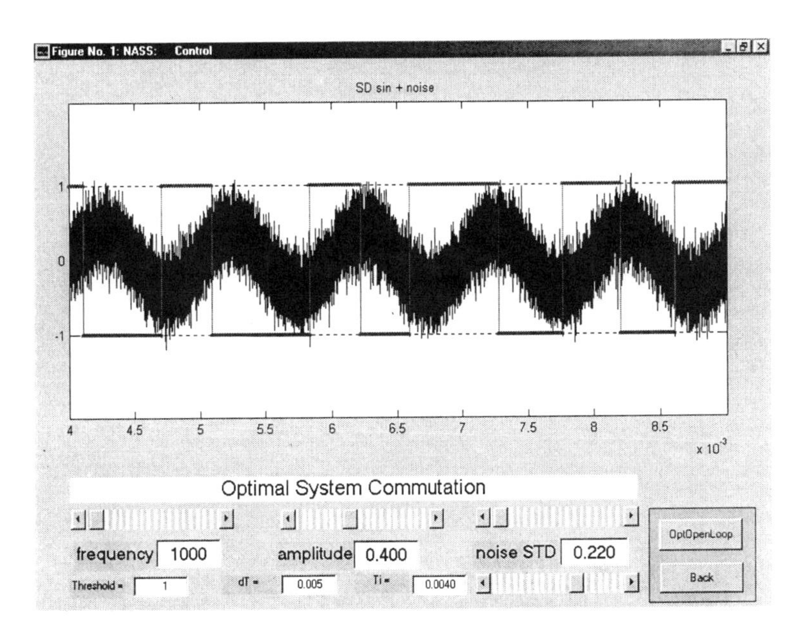

Figure 7. Result obtained by using the *Ampl and Freq* function. With a value of σ=0.220V the system switches.

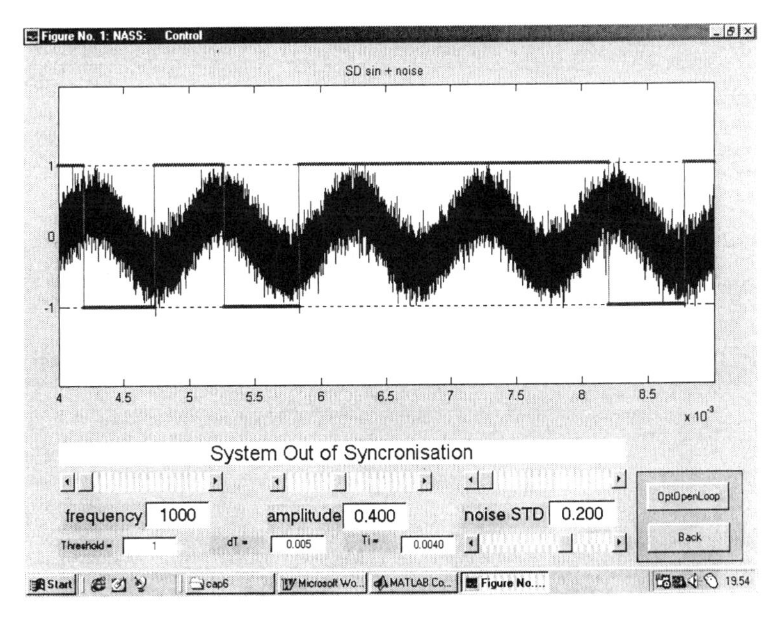

Figure 8. Effect of the noise standard deviation reduction. The system cannot follow the dynamic of the periodic forcing signal.

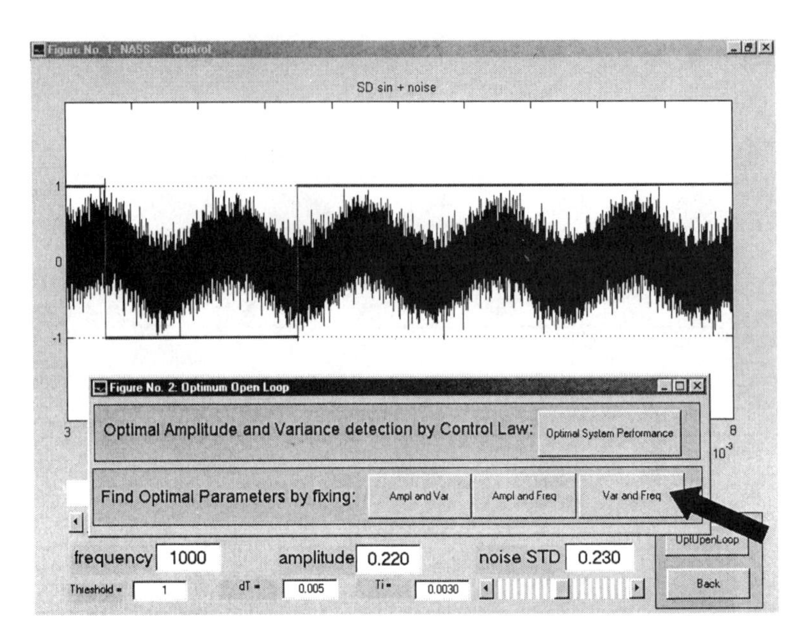

Figure 9. Tuning section for the detection of the minimum forcing signal amplitude. when the noise level has been altered (σ=0.230V).

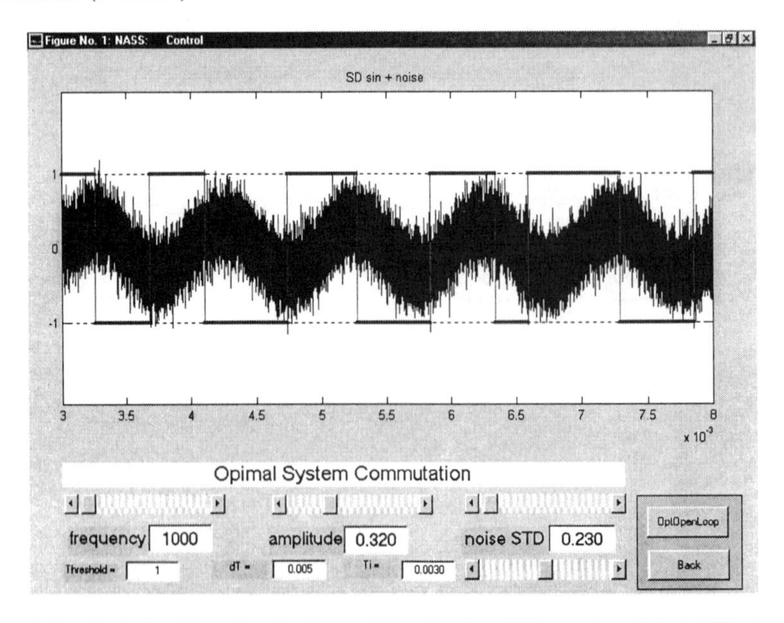

Figure 10. Result obtained by using the *Var and Freq* function. With a value of A=0.320V the system switches.

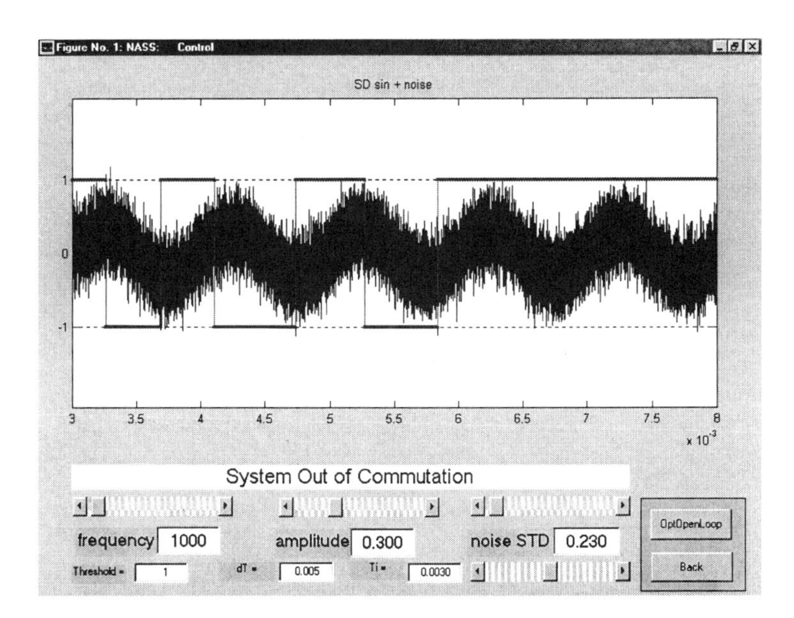

Figure 11. Effect of the forcing signal amplitude reduction. The system cannot follow the dynamic of the periodic forcing signal.

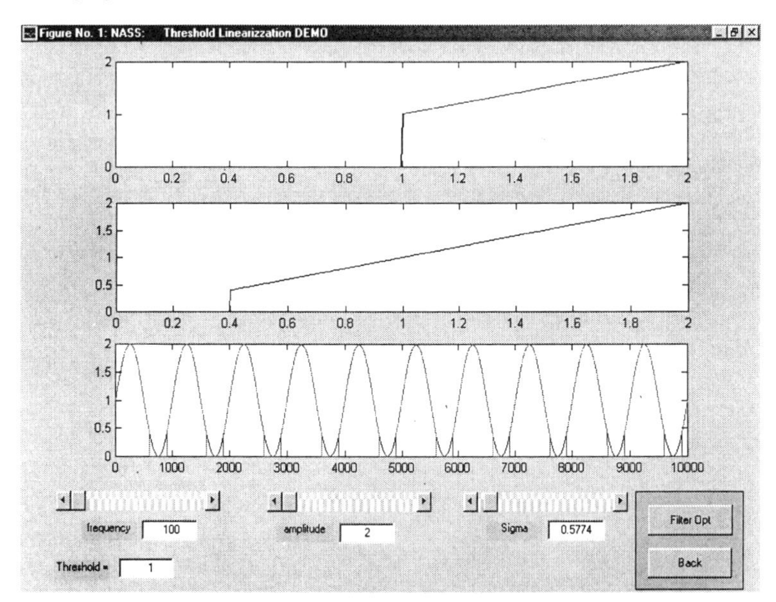

Figure 12. Effect of a low level of noise in case of linearisation task is required. The system is unable to process correctly input signals with an amplitude lower than the threshold.

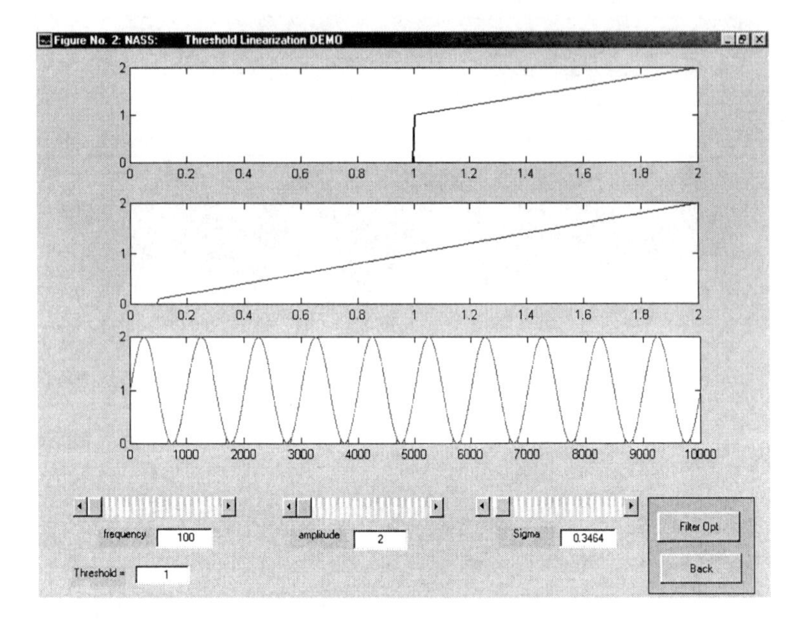

Figure 13. Result of optimal noise modulation allowing for linearisation task. By suitably modulating the level of the noise, the system made up of the original system and the filter is a linear one even when the input values are low.

References

[1] R. Benzi, A. Sutera and A. Vulpiani "The mechanism of Stochastic Resonance" J. Phys. A: Math. Gen.14L 453 (1981)

[2] P. Carbone, D. Petri, "Effect of Additive Dither on the Resolution of Ideal Quantisers",IEEE Trans. Instrum. Meas.; 1994, 43,3, 389-396.

APPENDIX A

APPENDIX A
BIBLIOGRAPHY ON NOISE ACTED SYSTEMS

Descriptive

[1] R. Benzi, A. Sutera and A. Vulpiani "The mechanism of Stochastic Resonance" J. Phys. A: Math. Gen.14L 453 (1981)

[2] J.P. Eckmann and L.E. Thomas "Remarks on stochastic resonance" J. Phys. A: Math. Gen. 15 L261 (1982)

[3] S. Fauve and F. Heslot, “Stochastic resonance in a bistable system" Phys. Lett. 97A 5 (1983)

[4] L. Gammaitoni F. Marchesoni, E. Menichella-Saetta and S. Santucci "Stochastic resonance in a bistable systems" Phys. Rev. Lett. 62 349 (1989)

[5] B. McNamara and K. Wiesenfeld "Theory of Stochastic Resonance" Phys. Rev. A39 4854 (1989)

[6] T.L. Carrol and L.M. Pecora "Stochastic Resonance and Crises" Phys. Rev. Lett. 70 576 (1993)

[7] M.I. Dykman, R. Mannella, P.V.E. McClintock and N.G. Stocks "Comment on Stochastic Resonance in bistable systems" Phys. Rev. Lett. 65, 2606 (1990)

[8] L. Gammaitoni, F. Marchesoni, E. Menichella-Saetta and S. Santucci "Reply to comment on: Stochastic Resonance in Bistable Systems" Phys. Rev. Lett. 65, 2607 (1990)

[9] L. Gammaitoni "Stochastic resonance in multi-threshold systems" Phys. Lett. A 208 315 (1995)

[10] Gammaitoni L. Hanggi P. Jung P. Marchesoni F. “Stochastic Resonance” [Review] Reviews of Modern Physics. 70(1):223-287, (1998)

[11] Dykman MI. Mcclintock PVE. “What Can Stochastic Resonance Do” Nature. 391(6665):344, (1998)

Theoretical

[12] R. Benzi A. Sutera and A. Vulpiani "Stochastic resonance in the Landau-Ginzburg equation" J. Phys A: Math. Gen. 18 2239 (1985)

[13] G. Debnath, T. Zhou and F. Moss "Remarks on stochastic resonance" Phys. Rev. 39A, 4323 (1989)

[14] R. F. Fox “Stochastic resonance in a double well" Phys. Rev. 39A, 4148 (1989)

[15] C. Presilla, F. Marchesoni and L. Gammaitoni "Periodically modulated bistable systems: non stationary statistical properties" Phys. Rev. 40A, 2105 (1989)

[16] L. Gammaitoni, E. Menichella-Saetta, S. Santucci, F. Marchesoni and C. Presilla "Periodically modulated bistable systems: Stochastic Resonance" Phys. Rev. 40A, 2114 (1989)

[17] L. Gammaitoni, F. Marchesoni, E. Menichella-Saetta, M. Punturo and S. Santucci "Stochastic Resonance: Phenomenology and applications" Non-Equilibrium Statistical Mechanics, World Scientific ed. (1989)

[18] P. Jung and P. Hanggi "Resonantly driven Brownian motion: Basic concepts and exact results" Phys. Rev. 41 A, 2977 (1990)

[19] T. Zhou, F. Moss, "Analog simulations of stochastic resonance" Phys. Rev. 41A, 4255 (1990)

[20] T. Zhou, F. Moss, P. Jung "Escape time distributions of a periodically modulated bistable system with noise" Phys. Rev. 42A, 3161 (1990)

[21] M.I. Dykman, P.V.E. McClintock, R. Mannella and N.G. Stocks "Stochastic resonance in the linear and nonlinear responses of a bistable system to a periodic field" Soviet Phys JETP Lett 52, 141 (1990)

[22] HU Gang, G. Nicolis and C. Nicolis "Periodically forced Fokker-Planck equation and stochastic resonance" Phys. Rev. 42A, 2030 (1990)

[23] U Gang, G. Nicolis and C. Nicolis "Comment on the ergodic properties of the periodically forced Fokker-Plank equation" Phys. Lett. A 151, 139 (1990)

[24] P. Jung and P. Hanggi "Amplification of small signals via stochastic resonance" Phys. Rev. A44 8032 (1991)

[25] P. Jung and P. Hanggi "Stochastische Resonanz" Physikalische B.tter 47: 1005 (1991)

[26] P. Jung, U. Behn, E. Pantazelou and F. Moss "Collective response in globally coupled bistable systems" Phys. Rev. A46 R1709-R1712 (1992)

[27] Simon and A. Libchaber "Escape and synchronization of a Brownian particle" Phys. Rev. Lett. 68 3375 (1992)

[28] M.I. Dykman, R. Mannella, P.V.E. McClintock, N.G. Stocks "Phase shifts in Stochastic Resonance" Phys. Rev. Lett. 68 2985 (1992)

[29] M. I. Dykman and P. V. E. McClintock "Power spectra of noise driven nonlinear systems and stochastic resonance" Physica D 58, 10 (1992)

[30] L. Gammaitoni, F. Marchesoni "Phase shifts in periodically modulated bistable potentials" Phys. Rev. Lett. 70 873 (1993)

[31] M.I. Dykman, R. mannella, P.V.E. McClintock, N.G. Stocks "Reply" Phys. Rev. Lett. 71 3625 (1993)

[32] P. Jung and P. Hanggi "Hopping and phase shifts in noisy periodically driven bistable systems" Z. Phys. B 90, 255 (1993)

[33] M. Morillo and J. Gomez-Ordonez "Phase Shifts in Driven Stochastic Nonlinear Systems" Phys. Rev. Lett. 71 9 (1993)

[34] L. Gammaitoni, F. Marchesoni, E. Menichella-Saetta and S. Santucci "Resonant crossing processes controlled by colored noise" Phys. Rev. Lett. 71 3625 (1993)

[35] E. Pantazelou, F. Moss, and D. Chialvo in *Noise in Physical Systems and 1/f Fluctuations*, edited by P. Handel and A. Chung (AIP Press, New York, 1993)

[36] L.Gammaitoni, F.Marchesoni, E.Menichella-Saetta and S.Santucci "Resonant crossings in bistable systems" in *Noise in Physical Systems and 1/f Fluctuations*, edited by P. Handel and A. Chung (AIP Press, New York, 1993)

[37] F. Moss, A. Bulsara and M. Shlesinger Proceedings of the NATO Advanced Research Workshop on Stochastic Resonance in Physics and Biology in J.Stat. Phys. 70, no.1/2 (1993)

[38] V.A. Shneidman, P. Jung and P. Hanggi "Power Spectrum of a driven bistable system" Europhys. Lett. 26: 571 (1994)

[39] R. F. Fox and Y. Lu "Analytic and numerical study of stochastic resonance" Phys. Rev. E48 3390 (1993)

[40] M I Dykman, H Haken, Gang Hu, D G Luchinsky, R Mannella, P V E McClintock C Z ing, N D Stein and N G Stocks "Linear response theory in stochastic resonance" Phys.Lett. A 180 332 (1993)

[41] Gang Hu, H. Haken, C.Z. Ning "Nonlinear-response effects in stochastic resonance" Phys.Rev. E47, 2321 (1993)

[42] P. Jung "Periodically driven stochastic systems" Phys.Rept. 234, 175 (1993)

[43] V.M. Makeyev "Stochastic resonance and its possible role in animate nature" Biophys 38, 189 (1993)

[44] G. Hu, T. Ditzinger, C.Z. Ning, H. Haken "Stochastic resonance without external periodic force" Phys. Rev. Lett. 71, 807 (1993)

[45] N.G. Stocks, N.D. Stein, P.V.E. McClintock "Stochastic resonance in monostable systems" J. Phys A26 L385 (1993)

[46] M I Dykman, D G Luchinsky, R Mannella, P V E McClintock, S M Soskin, N D Stein and N G Stocks "Stochastic resonance" Chap. 22 in Nonlinearity and Chaos in Engineering Dynamics ed. J M T Thompson and S R Bishop Wiley, New York, (1994)

[47] Bulsara, S. Lowen, and D. Rees "Cooperative behaviour in the periodically modulated Wiener process: Noise induced complexity in a model neuron" Phys. Rev. E 49, 4989 (1994)

[48] L. Gammaitoni, F. Marchesoni, E. Menichella-Saetta and S. Santucci "Multiplicative Stochastic Resonance" Phys. Rev. E49 4878 (1994)

[49] L. Gammaitoni, F. Marchesoni and S. Santucci “Stochastic Resonance without symmetry breaking" Phys. Lett. A 195, 116 (1994)

[50] F. Moss "Stochastic resonance: from the ice ages to the monkey's ear" in Some Contemporary Problems in Statistical Physics, edited by G. Weiss (SIAM, Philadelphia), 205 (1994)

[51] L.Gammaitoni, F.Marchesoni, E.Menichella-Saetta and S.Santucci "Stochastic Resonance in bistable systems with fluctuating barriers" Proceedings of the third Max Born Symposium, (1994)

[52] F. Moss, D. Pierson, D. O'Gorman, "Stochastic Resonance: Tutorial and update" Int. J. Bifurcation and Chaos 4(6) 1383. (1994)

[53] S. Vohra, L. Fabiny "Induced stochastic resonance near a subcritical bifurcation" Phys. Rev. E50 R2391 (1994)

[54] P. Jung "Threshold devices: fractal noise and neural talk" Phys. Rev. E50 2513 (1994)

[55] W.J. Rappel and S.H. Strogatz "Stochastic resonance in an autonomous system with a uniform limit cycle" Phys. Rev. E50 3249 (1994)

[56] M.C. Mahato and S.R. Shenoy "Hysteresis loss and stochastic resonance: A numerical study of a double-well potential" Phys. Rev. E 50, 2503 (1994)

[57] K. Wiesenfeld and F. Moss "Stochastic Resonance and the benefits of noise: from ice ages to crayfish and SQUIDs" Nature 373, 33 (1995)

[58] Kapitaniak T. "Mechanism of noise-induced resonance" Physical Review E. 52(1 Part B):1200-1201, (1995)

[59] Makarov DE. Makri N. "Stochastic resonance and nonlinear response in double-quantum-well structures" Physical Review B. 52(4):R2257-R2260, (1995)

[60] Hilgers, M Gremm, J. Schnakenberg "A criterion for stochastic resonance" Phys. Lett. A 209 313 (1995)

[61] F. Moss and K. Wiesenfeld "The Benefits of Background Noise" Scientific American Aug. (1995)

[62] J.J. Collins, C.C. Chow and T.T. Imhoff "Stochastic resonance without tuning" Nature 376, 236 (1995)

[63] A.J. Noest "Tuning Stochastic Resonance" Nature 378, 341 (1995)

[64] Collins JJ. Chow CC. Imhoff TT. "Tuning Stochastic Resonance - reply" Nature. 378(6555):341-342, (1995)

[65] Wackerbauer R. "Noise-induced stabilization of the Lorentz system" Physical Review E. 52(5 Part A):4745-4749, (1995)

[66] L. Gammaitoni "Stochastic resonance and the dithering effect in threshold physical systems" Phys. Rev. E52 4691 (1995)

[67] Brey JJ. Casadopascual J. Sanchez B. "Resonant Behaviour of a poisson process driven by a periodic signal" Physical Review E. 52(6 Part A):6071-6081, (1995)

[68] Berghaus C. Hilgers A. Schnakenberg J. "Monte Carlo And Numerical Studies Of Coloured Noise And Stochastic Resonance Problems" Zeitschrift fur Physik B-Condensed Matter. 100(1):157-163, (1996)

[69] Gang H. Daffertshofer A. Haken H. "Diffusion Of Periodically Forced Brownian Particles Moving In Space-Periodic Potentials" Physical Review Letters. 76(26):4874-4877, (1996)

[70] R. Bulsara, L. Gammaitoni "Tuning in to Noise" Physics Today March p. 39, (1996)

[71] Chattah AK. Briozzo CB. Osenda O. Caceres MO. "Signal-To-Noise Ratio In Stochastic Resonance" Modern Physics Letters B. 10(22):1085-1094, (1996)

[72] Drozdov AN. Morillo M. "Validity Of Basic Concepts In Nonlinear Cooperative Fokker-Planck Models" Physical Review E. 54(4 Part A):3304-3313, (1996)

[73] Moss F. Chioutan F. Klinke R. "Will There Be Noise In Their Ears" Nature Medicine. 2(8):860-862, (1996)

[74] Berdichevsky V. Gitterman M. "Multiplicative Stochastic Resonance In Linear Systems - Analytical Solution" Europhysics Letters. 36(3):161-165, (1996)

[75] Loerincz K. Gingl Z. Kiss LB. "A Stochastic Resonator Is Able To Greatly Improve Signal-To-Noise Ratio" Physics Letters A. 224(1-2):63-67, (1996)

[76] Berdichevsky V. Gitterman M. "Stochastic Resonance In A Bistable Piecewise Potential - Analytical Solution" Journal of Physics A-Mathematical & General. 29(18):L 447-L 452, (1996)

[77] Lanzara E. Mantegna RN. Spagnolo B. Zangara R. "Experimental Study Of A Nonlinear System In The Presence Of Noise - The Stochastic Resonance" American Journal of Physics. 65(4):341-349, (1997)

[78] Bezrukov SM. Vodyanoy I. "Stochastic Resonance In Non-Dynamical Systems Without Response Thresholds" Nature. 385(6614):319-321, (1997)

[79] Jung P. Wiesenfeld K. "Too Quiet To Hear A Whisper" Nature. 385(6614):291, (1997)

[80] Vilar JMG. Rubi JM. "Stochastic Multiresonance" Physical Review Letters. 78(15), (1997)

[81] Dhara AK. "Enhancement Of Signal-To-Noise Ratio" Journal of Statistical Physics. 87, (1997)

[82] Barzykin AV., Seki K. "Stochastic Resonance Driven By Gaussian Multiplicative Noise" Europhysics Letters. 40(2), (1997)

[83] Mahato MC. Jayannavar AM. "Some Stochastic Phenomena In A Driven Double-Well System" Physica A. 248(1-2):138-154, (1998)

[84] Tretyakov MV. "Numerical Technique For Studying Stochastic Resonance" Physical Review A. 57(4):4789-4794, (1998)

[85] V. S. Anishchenko, M. A. Safonova, L.O. Chua "Stochastic resonance in Chua's circuit " Int. J. Bifurcation and Chaos 2(2), 397-401, (1992)

[86] V. S. Anishchenko, M. A. Safonova, L.O. Chua "Stochastic resonance in the nonantonomous Chua's Circuit " J. Cir. .Sys. and Cmp. 3, 553-578, (1993)

[87] V. S. Anishchenko, M. A. Safonova, L.O. Chua "Stochastic resonance in Chua's circuit driven by amplitude or frequency modulated signals" Int. J. Bifurcation and Chaos 4(2) 441-446, (1994)

[88] Crisanti, M. Falcioni, G. Paladin and A. Vulpiani "Stochastic resonance in deterministic chaotic systems" J. Phys. A: Math. Gen. 27 L597 (1994)

[89] Longtin "Mechanisms of stochastic phase locking" CHAOS 5, 209-215 (1995)

[90] M. Inchiosa, A. Bulsara, J. Lindner, B. Meadows, and W. Ditto "Array Enhanced Stochastic Resonance: Implications for Signal Processing" Proceedings of the Third Technical Conference on Nonlinear Dynamics (Chaos) and Full Spectrum Processing (1995)

[91] Kim S. Reichl LE. "Stochastic Chaos And Resonance In A Bistable Stochastic System" Physical Review E. 53(4 Part A):3088-3095, (1996)

[92] Franaszek M. Simiu E. "Stochastic Resonance - A Chaotic Dynamics Approach" Physical Review E. 54(2):1298-1304, (1996)

[93] Khovanov IA. Anishchenko VS. "Mechanism Of Stochastic Resonance In The Chaos-Chaos Alternation System" [Russian] Pisma V Zhurnal Tekhnicheskoi Fiziki. 22(20):75-81, (1996)

[94] Roy M. Amritkar "Re Effect Of Noise On Coupled Chaotic Systems" Pramana-Journal of Physics. 48(1):271-285, (1997)

[95] Gassmann F. "Noise-Induced Chaos-Order Transitions" Physical Review E. 55(3 Part A):2215-2221, (1997)

[96] Reibold E. Just W. Becker J. Benner H. "Stochastic Resonance In Chaotic Spin-Wave Dynamics" Physical Review Letters. 78(16):3101-3104, (1997)

[97] Osipov VV. Ponizovskaya EV. "The Nature Of Bursting Noises, Stochastic Resonance And Deterministic Chaos In Excitable Neurons" Physics Letters A. 238(6):369-374, (1998)

[98] Stone L. Saparin PI. Huppert A. Price C. EL Nino "Chaos - The Role Of Noise And Stochastic Resonance On The Enso Cycle" Geophysical Research Letters. 25(2):175-178, (1998)

[99] B. Andò, S. Baglio, S. Graziani, N. Pitrone, A Probabilistic Approach to the Threshold Error Reduction Theory in Bistable Measurement Devices, IMTC'98, St. Paul Minnesota, USA May 18-21, (1998)

Applications

[100] R. Benzi, G. Parisi, A. Sutera and A. Vulpiani "Stochastic resonance in climatic change" Tellus, 34 10 (1982)

[101] C.Nicolis "Stochastic aspects of climatic transitions - response to a periodic forcing" Tellus 34 1 (1982)

[102] R. Benzi, G. Parisi, A. Sutera and A. Vulpiani "A theory of stochastic resonance in climatic change" SIAM J. Appl Math 43, 565 (1983)

[103] McNamara K. Wiesenfeld and R. Roy "Observation of Stochastic Resonance in a ring laser" Phys. Rev. Lett. 60 2626 (1988)

[104] P. Jung and P. Hanggi "Stochastic nonlinear dynamics modulated by external periodic forces" Europhys. Lett. 8 505 (1989)

[105] G. Matteucci "Orbital forcing in a stochastic resonance model of the late Pleistocene climatic variations " Climate Dyn. 3, 179 (1989)

[106] G. Vemuri and R. Roy "Stochastic resonance in a bistable ring laser" Phys. Rev. 39A, 4668 (1989)

[107] P. Jung "Thermal activation in bistable systems under external periodic forces" Z. Phys. B 76, 521 (1989)

[108] L.Gammaitoni, S.Santucci, M.Giordano, M.Martinelli and L.Pardi "Bistability and Stochastic Resonance in Electron Paramagnetic Resonance" Proc. 25 Congress Ampere p. 422, Stuttgart (1990)

[109] A.N. Grigorenko, V.I. Konov and P.I. Nikitin "Magnetostochastic resonance" Pis'ma Zh. Eksp. Teor. Fiz. 52, 1182 (1990) and JETP Lett., 52, 593 (Dec 1990)

[110] M. I. Dykman, A.L. Velikovich, G.P. Golubev, D.G. Luchinskii and S.V. Tsuprikov "Stochastic resonance in an all-optical passive bistable system" Soviet Phys JETP Lett 53, 193 (1991)

[111] F. Moss Ber Bunsenges. Phys. Chem. 95 No. 3 (1991)

[112] P. Jung, U. Behn, E. Pantazelou and F. Moss "Collective response in globally coupled bistable systems" Phys. Rev. A 46, R1709 (1991)

[113] Maddox "Toward the brain-computer's code" Nature 352, 469 (1991)

[114] Bulsara, E. Jacobs, T. Zhou, F. Moss and L. Kiss "Stochastic resonance in a single neuron model: theory and analog simulation" J. Theor. Biol. 152, 531 (1991)

[115] L. Gammaitoni, M. Martinelli, L. Pardi and S. Santucci "Observation of Stocastic Resonance in EPR Systems" Phys. Rev. Lett. 67, 1799 (1991)

[116] L. Gammaitoni, F. Marchesoni, M. Martinelli, L. Pardi and S. Santucci "Phase shifts in bistable EPR systems at stochastic resonance" Phys. Lett. A 158 449 (1991)

[117] Grossmann,T. Dittrich,P. Jung, and P. Hanggi "Coherent Destruction of Tunneling" Phys. Rev. Lett. 67: 516 (1991).

[118] G. Matteucci "A study of the climatic regimes of the Pleistocene using a stochastic resonance model" Climate Dyn. 6, 67 (1991).

[119] Maddox Nature 352 469 (1991)

[120] Longtin, A. Bulsara, and F. Moss "Time interval sequences in the bistable systems and the noise induced transmission of information by sensory neurons" Phys. Rev. Lett. 67, 656 (1991)

[121] E. K. Sadykov "Stochastic resonance in ultra-despersed magnetics. neurons" Fiz. Tv. Tela, 33, No.11, 3302-3307 (1991)

[122] M.L. Spano and M. Wun-Fogle, W.L. Ditto "Experimental observation of stochastic resonance in a magnetoelastic ribbon" Phys. Rev. A46 R5253 (1992)

[123] 47.M. I. Dykman, D. G. Luchinsky, P. V. E. McClintock, N. D. Stein and N. G. Stocks "Stochastic resonance for periodically modulated noise intensity" Phys. Rev. A46 R 1713 (1992)

[124] L. Gammaitoni, M. Martinelli, L. Pardi and S. Santucci "Noise induced phenomena in Electron Paramagnetic Resonance Systems" Mod. Phys. Lett. B 6 197 - (1992)

[125] D.S. Leonard "Stochastic resonance in a random walk" Phys. Rev. A46 6742 (1992)

[126] N G Stocks, N D Stein, S M Soskin and P V E McClintock "Zero-dispersion stochastic resonance" J Phys A 25, L1119 (1992)

[127] Gang Hu, H. Haken, C.Z. Ning "A study of stochastic resonance without adiabatic approximation" Phys.Lett. A 172, 21 (1992)

[128] Ch. Zerbe, P. Jung and P. Hanggi "Lasers with Injected Signals: Fluctuations and Linewidth" Z. Physik B86:151 (1992)

[129] E. K. Sadykov "Stochastic resonance in small-particle magnetics: I. Radiospectroscopic study. " J. Phys.: Condens. Matter, 4, 3295 (1992)

[130] Bulsara and G. Schmera "Stochastic resonance in globally coupled interacting oscillators" Phys. Rev. E 47, 3734 (1993)

[131] A.Bulsara, A. Maren, and G. Schmera Biol. Cyb. 70, 145 (1993)

[132] Bulsara and A. Maren in Rethinking Neural Networks: Quantum Fields and Biological Data,edited by K. H. Pribram (L. Erlbaum Associates, 1993)

[133] Hibbs, E. Jacobs, J. Bekkedahl, A. Bulsara, and F. Moss "Stochastic resonance in a bistable SQUID loop" in Noise in Physical Systems and 1/f Fluctuations, edited by P. Handel and A. Chung (AIP Press, New York, 1993)

[134] M I Dykman, D G Luchinsky, R Mannella, P V E McClintock, N D Stein and N G Stocks "High frequency stochastic resonance in periodically driven systems" Soviet Phys JETP Lett 58, 150 (1993) (from Pis'ma Zh. Eksp. Teor. Fiz. 58, 145 (1993)

[135] T. Albert, A. Bulsara, G. Schmera, and M. Inchiosa in Conference Record of the Twenty-Seventh Asilomar Conference on Signals, Systems and Computers, edited by A. Singh (IEEE Society Press, Los Alamitos, CA, 1993)

[136] F. Grossmann, P. Hanggi, "AC-driven quantum decay" J. Chem. Phys. 170, 295 (1993)

[137] J. Douglass, L. Wilkens, E. Pantazelou, and F. Moss "Noise enhancement of the information transfer in crayfish mechanoreceptors by stochastic resonance" Nature 365 337 (1993)

[138] G. De-chun, H. Gang, W. Xia-dong, Y. Chun-yan, Q.Guang-rong, L. Rong, and D. Da-fu Phys.Rev. E 8,4862 (1993)

[139] K. Wiesenfeld, D. Pierson, E. Pantazelou, and F. Moss "Stochastic resonance on a circle" Phys. Rev. Lett. 72, 2125 (1994)

[140] R. Bartussek, P. Hanggi, P. Jung "Stochastic Resonance In Optical Bistable Systems" Phys. Rev. E49 3930 (1994)

[141] D.S. Leonard, L.E. Reichl "Stochastic resonance in a chemical reaction" Phys. Rev. E49 1734 (1994)

[142] M I Dykman, D G Luchinsky, R Mannella, P V E McClintock, N D Stein and N G Stocks "Supernarrow spectral peaks and high-frequency stochastic resonance in systems with periodic attractors" Phys Rev E 49, 1198 (1994)

[143] R.N. Mantegna, B. Spagnolo "Stochastic resonance in a tunnel diode" Phys. Rev. E49 R1792 (1994)

[144] M. I. Dykman, D. G. Luchinsky, and R. Mannella in Fluctuations and Order: The New Synthesis edited by M. Millonos (Springer Verlag, New York, 1994)

[145] J. Maddox "Bringing more order out of noisiness" Nature 369, 271 (1994)

[146] Longtin, A. Bulsara, D. Pierson, and F. Moss "Bistability and the dynamics of periodically forced sensory neurons" Biol. Cybern. 70, 569 (1994)

[147] R. Lofstedt, S.N. Coppersmith "Quantum stochastic resonance" Phys. Rev. Lett. 72 1947 (1994)

[148] M. Riani and E. Simonotto "Stochastic resonance in the perceptual interpretation of ambiguous figures, a neural network model" Phys. Rev. Lett. 72 3120 (1994)

[149] V.A. Shneidman, P. Jung and P. Hanggi "Weak-noise limit of stochastic resonance" Phys. Rev. Lett. 72 2682 (1994)

[150] M I Dykman, G P Golubev, D G Luchinsky, P V E McClintock, N D Stein and N G Stocks "Noise-enhanced heterodyning in bistable systems" Phys Rev E 49, 1935 (1994)

[151] N. Grigorenko, P. I. Nikitin, A. N. Slavin, and P. Y. Zhou "Experimental observation of magnetostochastic resonance." J. Appl. Phys. 76, No.10, 6335 (1994)

[152] Yu. L. Raikher and V. I. Stepanov "Stochastic resonance in single-domain particles." J. Phys.: Condens. Matter, 6, 4137 (1994)

[153] Hibbs, A.L. Singsaas, E.W. Jacobs, A. Bulsara, J. Bekkedahl "Stochastic resonance in a superconducting loop with a Josephson junction" J. Appl. Phys. 77 2582 (1995)

[154] J.M. Casado, J.J. Mejias, M. Morillo "Comments on 'Stochastic resonance in a periodic potential system under a constant driving force" Phys. Lett. 197A 365 (1995)

[155] P. Jung, G. Mayer-Kress "Spatio-Temporal Stochastic Resonance in Excitable Media"

[156] Jung P. Talkner P. "Suppression of higher harmonics at noise induced resonances" Physical Review E. 51(3 Part B):2640-2643, (1995)

[157] Reale DO. Pattanayak AK. Schieve WC. "Semiquantal corrections to stochastic resonance" Physical Review E. 51(4 Part A):2925-2932, (1995)

[158] R. Rouse, S. Han and J.E. Lukens "Flux amplification using stochstic superconducting quantum interference devices" Appl. Phys. Lett. 66 108 (1995)

[159] Gammaitoni L. Marchesoni F. Menichella-Saetta E. Santucci S. "Stochastic Resonance In The Strong-Forcing Limit" Physical Review E. 51(5 Part A):R3799-R3802, (1995)

[160] Dykman MI. Horita T. Ross J. "Statistical distribution and stochastic resonance in a periodically driven chemical system" Journal of Chemical Physics. 103(3):966-972, (1995)

[161] Morillo M. Gomezordonez J. Casado JM. "Stochastic resonance in a mean-field model of cooperative behaviour" Physical Review E. 52(1 Part A):316-320, (1995)

[162] Z. Gingl, L.B. Kiss and F. Moss "Non-Dynamical Stochastic Resonance: Theory and Experiments with White and Arbitrarily Coloured Noise" Europhys. Lett. 29 191 (1995)

[163] Raikher YL. Stepanov VI. "Stochastic resonance and phase shifts in superparamagnetic particles" Physical Review B-Condensed Matter. 52(5):3493-3498, (1995)

[164] Casado JM. Morillo M. "Langevin description of the response of a stochastic mean-field model driven by a time-periodic field" Physical Review E. 52(2):2088-2090, 1995 Aug. Phys. Rev. Lett. 74(11), 2130 (1995)

[165] Li R. Qin GR. Hu G. Wen XD. Zhu HJ. "Signal-to-Noise in bistable systems subject to signal and monochromatic noise" Communications in Theoretical Physics. 24(1):19-26, (1995)

[166] R. Li, G. Hu, C. Yang, X. Weng, G. Qing, and H. Zhu "Stochastic resonance in bistable systems subject to signal and quasimonochromatic noise" Phys. Rev. E51 3964 (1995)

[167] J. Lindner, B. Meadows, W. Ditto, M. Inchiosa and A. Bulsara, "Array Enhanced Stochastic Resonance and Spatiotemporal synchronization" Phys. Rev. Lett. 75 3 (1995)

[168] Neiman and L. Schimansky-Geier Phys.Lett. 197 A 397 (1995)

[169] L. Gammaitoni, F. Marchesoni and S. Santucci "Stochastic Resonance as a Bona Fide resonance" Phys. Rev. Lett. 74 1052 (1995)

[170] P. Jung "Stochastic Resonance and optimal design of threshold detectors" Phys. Lett. A 207, 93 (1995)

[171] Grigorenko AN. Nikitin PI. "Stochastic resonance in a bistable magnetic system" IEEE Transactions on Magnetics. 31(5):2491-2493, (1995)

[172] Phillips JC. Schulten K. "Diffusive hysteresis at high and low driving frequencies" Physical Review E. 52(3 Part A):2473-2477, (1995)

[173] Stemmler M. Usher M. Niebur E. "Lateral interactions in primary visiual cortex - A model bridging physiology and psychophysics" Science. 269(5232):1877-1880, (1995)

[174] F. Moss, Xing Pei "Neurons in parallel" Nature 376, 211 (1995)

[175] Lin I and Jeng-Mei Liu "Experimental Observation of Stochastic Resonancelike Behaviour of Autonomous Motion in Weakly Ionized rf Magnetoplasms" Phys. Rev. Lett. 74 3161 (1995)

[176] Collection of papers presented at the International Workshop Fluctuation in Physics and Biology: Stochastic Resonance, Signal Processing and Related Phenomena Elba, 5-10 June 1994 Nuovo Cimento 17D 7-8 (1995)

[177] Perez-Madrid and J.M. Rubi "Stochastic Resonance in a system of ferromagnetic particles" Phys. Rev. E51 4159 (1995)

[178] Pei X. Bachmann K. Moss F. "The detection threshold, noise and stochastic resonance in the Fitzhugh-Nagumo neuron model" Physics Letters A. 206(1-2):61-65, (1995)

[179] Grifoni M. Sassetti M. Hanggi P. Weiss U. "Cooperative effects in the nonlinearly driven spin-boson system" Physical Review E. 52(4 Part A):3596-3607, (1995)

[180] Fulinski A. "Relaxation, noise-induced transitions, and stochastic resonance driven by non-markovian dichotomic noise" Physical Review E. 52(4 Part B):4523-4526, (1995)

[181] J. Masoliver, A. Robinson, G. H. Weiss "Coherent stochastic resonance" Phys. Rev. Lett. 74 1052 (1995)

[182] W. Yang, M. Ding, Hu Gang "Trajectory (Phase) Selection in Multistable System: Stochastic Resonance, Signal Bias, and the Effect of Signal Phase" Phys. Rev. Lett. 74 3955 (1995)

[183] S.M. Bezrukov, I. Vodyanoy "Noise-induced enhancement of signal transduction across voltage-dependent ion channels" Nature 378, 362 (1995)

[184] M I Dykman, G P Golubev, I Kh Kaufman, D G Luchinsky, P V E McClintock, and E A Zhukov "Noise-enhanced optical heterodyning in an all-optical system" Appl Phys Lett 67, 308 (1995)

[185] Gitterman M. Weiss GH. "Coherent stochastic resonance in the presence of a field" Physical Review E. 52(5 Part B):5708-5711, (1995)

[186] Bulsara AR. Lowen SB. Rees CD. "Coherent stochastic resonance in the presence of a field - reply" Physical Review E. 52(5 Part B):5712-5713, (1995)

[187] Z. Néda, "Stochastic resonance in Ising systems" Phys. Rev. E51 5315 (1995)

[188] M. E. Inchiosa and A. R. Bulsara "Nonlinear dynamic elements with noisy sinusoidal forcing: enhancing response via nonlinear coupling" Physical Review E, vol. 52, pp. 327-339 (1995)

[189] J.J. Collins, C.C. Chow and T.T. Imhoff "Aperiodic stochastic resonance in excitable systems" Phys. Rev. E52 R3321 (1995)

[190] Shulgin, A. Neiman, V. Anishchenko "Mean Switching Frequency Loking in Stochastic Bistable System Driven by a Periodic Force" Phys. Rev. Lett. 75 4157 (1995)

[191] Babinec P. Babincova M. "Noise-enhanced electroweak bioenantiosection in the Frank model" Ach-Models in Chemistry. 132(5):853-857, (1995)

[192] Makarov DE. Makri N. "Control of dissipative tunneling dynamics by continuous wave electromagnetic fields-localization and large amplitude coherent motion" Physical Review E 52(6 Part A):5863-5872, (1995)

[193] M. E. Inchiosa and A. R. Bulsara, "Coupling Enhances Stochastic Resonance in Nonlinear Dynamic Elements Driven by a Sinusoid Plus Noise" Physics Letters A, vol. 200, pp. 283-288 (1995)

[194] "Stochastic Resonance in 3D Ising ferromagnets" Physics Letters A 210(1-2):125-128, (1996)

[195] M. E. Inchiosa and A. R. Bulsara "Signal Detection Statistics of Stochastic Resonators" Physical Review E, vol. 53, pp. 2021R-2024R (1996)

[196] Chapeaublondeau F. Godivier X. Chambet N. "Stochastic Resonance in a neuron model that transmits spike trains" Physical Review E 53(1 Part B):1273-1275, (1996)

[197] Mantegna RN. Spagnolo B. "Noise enhanced stability in an unstable system" Physical Review Letters. 76(4):563-566, (1996)

[198] J. F. Lindner, B. K. Meadows, W. L. Ditto, M. E. Inchiosa, and A. R. Bulsara "Scaling Laws for Spatiotemporal Synchronization and Array Enhanced Stochastic Resonance" Physical Review E, vol. 53, pp. 2081-2086 (1996)

[199] Jost BM. Saleh BEA "Signal-to-Noise ratio improvement by stochastic resonance in a unidirectional photorefractive ring resonator" Optics Letters. 21(4):287-289, (1996)

[200] Longtin and K. Hinzer " Encoding with bursting, subthreshold oscillations and noise in mammalian cold receptors" NEURAL COMPUT. 8, 215-255 (1996)

[201] Grigorenko AN. Nikitin PI. Magnetostochastic Resonance As A New Method For Investigations Of Surface And Thin Film Magnetism" Applied Surface Science. 92:466-470, (1996)

[202] Grifoni M. Hanggi P. "Coherent and incoherent quantum stochastic resonance" Phys. Rev. Lett. 76(10):1611-1614, (1996)

[203] Levin JE., Miller JP. "Broadband neural encoding in the cricket cercal sensory system enhanced by stochastic resonance" Nature. 380(6570):165-168, (1996)

[204] Loreto V. Paladin G. Vulpiani A. "Concept Of Complexity In Random Dynamical Systems" Physical Review E. 53(3):2087-2098, (1996)

[205] F. Marchesoni, L. Gammaitoni, A.R. Bulsara "Spatiotemporal stochastic resonance in a phi4 model of kink-antikink nucleation" Phys. Rev. Lett. 76:2609-2612, (1996)

[206] Hohmann W. Muller J. Schneider FW. "Stochastic Resonance In Chemistry .3. The Minimal-Bromate Reaction" Journal of Physical Chemistry. 100(13):5388-5392, (1996)

[207] Grigorenko AN. Nikitin PI. Roschepkin GV. "Frequency Mixing Phenomena In A Bistable System" Journal of Applied Physics. 79(8 Part 2B):6113-6115, (1996)

[208] Bulsara AR. Elston TC. Doering CR. Lowen SB. Lindenberg K. "Cooperative Behavior In Periodically Driven Noisy Integrate-Fire Models Of Neuronal Dynamics" Physical Review E. 53(4 Part B):3958-3969, (1996)

[209] Adair RK. "Didactic Discussion Of Stochastic Resonance Effects And Weak Signals" Bioelectromagnetics. 17(3):242-245, (1996)

[210] Kaplan DT. Clay JR. Manning T. Glass L. Guevara MR. Shrier A. "Subthreshold Dynamics In Periodically Stimulated Squid Giant Axons" Physical Review Letters. 76(21):4074-4077, (1996)

[211] Chapeaublondeau F. "Stochastic Resonance In The Heaviside Nonlinearity With White Noise And Arbitrary Periodic Signal" Physical Review E. 53(5 Part B):5469-5472, (1996)

[212] Neiman A. Shulgin B. Anishchenko V. Ebeling W. Schimanskygeier L. Freund J. "Dynamical Entropies Applied To Stochastic Resonance" Physical Review Letters. 76(23):4299-4302, (1996)

[213] Brey JJ. Prados A. "Stochastic Resonance In A One-Dimensional Ising Model" Physics Letters A. 216(6):240-246, (1996)

[214] Grifoni M. Hartmann L. Berchtold S. Hanggi P. "Quantum Tunneling And Stochastic Resonance" Physical Review E. 53(6 Part A):5890-5898, (1996)

[215] Collins JJ. Imhoff TT. Grigg P. "Noise-Enhanced Information Transmission In Rat Sa1 Cutaneous Mechanoreceptors Via Aperiodic Stochastic Resonance" Journal of Neurophysiology. 76(1):642-645, (1996)

[216] Sadykov EK. Isavnin AG. "Theory Of Dynamic Magnetic Susceptibility Of Uniaxial Superparamagnetic Particles" [Russian] Fizika Tverdogo Tela. 38(7):2104-2112, (1996)

[217] Alibegov MM. "Stochastic Resonance, The Rayleigh Test, And Identification Of The 25-Day Periodicity In The Solar Activity Astronomy" Letters-A Journal of Astronomy & Space Astrophysics. 22(4):564-572, (1996)

[218] Hafemeister D. RESOURCE LETTER BELFEF-1 "Biological Effects Of Low-Frequency Electromagnetic Fields" [Review] American Journal of Physics. 64(8):974-981, (1996)

[219] Grifoni M. Hanggi P. "Nonlinear Quantum Stochastic Resonance" Physical Review E. 54(2):1390-1401, (1996)

[220] Morse RP. Evans EF. "Enhancement Of Vowel Coding For Cochlear Implants By Addition Of Noise" Nature Medicine. 2(8):928-932, (1996)

[221] Godivier X. Chapeaublondeau F. "Noise-Enhanced Transmission Of Spike Trains In The Neuron" Europhysics Letters. 35(6):473-477, (1996)

[222] Gang H. Haken H. Fagen X. "Stochastic Resonance With Sensitive Frequency Dependence In Globally Coupled Continuous Systems" Physical Review Letters. 77(10):1925-1928, (1996)

[223] Kaufman IK. Luchinsky DG. Mcclintock PVE. Soskin SM. Stein ND. "High-Frequency Stochastic Resonance In Squids" Physics Letters A. 220(4-5):219-223, (1996)

[224] Bulsara AR. Inchiosa ME. Gammaitoni L. "Noise-Controlled Resonance Behavior In Nonlinear Dynamical Systems With Broken Symmetry" Physical Review Letters. 77(11):2162-2165, (1996)

[225] Chattah AK. Briozzo CB. Osenda O. Caceres MO. "Effect Of Thermal Noise On The Current-Voltage Characteristics Of Josephson Junctions Modern" Physics Letters B. 10(22), (1996)

[226] Vilar JMG. Rubi JM. "Divergent Signal-To-Noise Ratio And Stochastic Resonance In Monostable Systems" Physical Review Letters. 77(14):2863-2866, (1996)

[227] Gaveau B. Moreau M. Danielak R. Frankowicz M. "Exact Results For The Transmission Probabilities In Linear Array Of Fluctuating Barriers" Acta Physica Polonica B. 27(9) (1996)

[228] Bulsara AR. Zador A. "Threshold Detection Of Wideband Signals - A Noise-Induced Maximum In The Mutual Information" Physical Review E. 54(3):R2185-R2188, (1996)

[229] Heneghan C. Chow CC. Collins JJ. Imhoff TT. Lowen SB. Teich MC. "Information Measures Quantifying Aperiodic Stochastic Resonance" Physical Review E. 54(3):R2228-R2231, (1996)

[230] Dykman MI. Luchinsky DG. Mannella R. Mcclintock PVE. Soskin SM. Stein ND. "Resonant Subharmonic Absorption And Second-Harmonic Generation By A Fluctuating Nonlinear Oscillator" Physical Review E. 54(3):2366-2377, (1996)

[231] Rappel WJ. Karma A. "Noise-Induced Coherence In Neural Networks" Physical Review Letters. 77(15):3256-3259, (1996)

[232] Cordo P. Inglis JT. Verschueren S. Collins JJ. Merfeld DM. Rosenblum S. Buckley S. Moss F. "Noise In Human Muscle Spindles" Nature. 383(6603):769-770, (1996)

[233] Collins JJ. Imhoff TT. Grigg P. "Noise-Enhanced Tactile Sensation" Nature. 383 (1996)

[234] Grifoni M. "Dynamics Of The Dissipative Two-State System Under Ac Modulation Of Bias And Coupling Energy" Physical Review E. 54(4 Part A):R3086-R3089, (1996)

[235] Wio HS. "Stochastic Resonance In A Spatially Extended System" Physical Review E. 54(4 Part A):R3075-R3078, (1996)

[236] Dubinov AE. Mikheev KE. Nizhegorodtsev YB. Selemir VD. "On The Stochastic Resonance In Ferroelectrics" [Russian] Izvestiya Akademii Nauk Seriya Fizicheskaya. 60(10):76-77, (1996)

[237] Collins JJ. Chow CC. Capela AC. Imhoff TT. "Aperiodic Stochastic Resonance" Physical Review A. 54(5):5575-5584, (1996)

[238] Gluckman BJ. Netoff TI. Neel EJ. Ditto WL. Spano ML. Schiff SJ. "Stochastic Resonance In A Neuronal Network From Mammalian Brain" Source Physical Review Letters. 77(19) (1996)

[239] Salman H. Soen Y. Braun E. "Voltage Fluctuations And Collective Effects In Ion-Channel Protein Ensembles" Physical Review Letters. 77(21):4458-4461, (1996)

[240] Pei X. Wilkens L. Moss F. "Noise-Mediated Spike Timing Precision From Aperiodic Stimuli In An Array Of Hodgekin-Huxley-Type Neurons" Physical Review Letters. 77(22):4679-4682, (1996)

[241] Burlak GN. Ishkabulov K. Mironenko SO. "Stochastic Resonance In The Soliton Dynamics System" [Russian] Pisma V Zhurnal Tekhnicheskoi Fiziki. 22(22):66-71, (1996)

[242] Locher M. Johnson GA. Hunt ER. "Spatiotemporal Stochastic Resonance In A System Of Coupled Diode Resonators" Physical Review Letters. 77(23):4698-4701, (1996)

[243] Neiman A. Shulgin B. Anishchenko V. Ebeling W. Schimanskygeier L. Freund J. "Dynamical Entropies Applied To Stochastic Resonance" (VOL 76, PG 4299, 1996) Physical Review Letters. 77(23):4851, (1996)

[244] Maier RS. Stein DL. "Oscillatory Behavior Of The Rate Of Escape Through An Unstable Limit Cycle" Physical Review Letters. 77(24):4860-4863, (1996)

[245] Neiman A. Sung W. "Memory Effects On Stochastic Resonance" Physics Letters A. 223(5):341-347, (1996)

[246] Yang HL. Huang ZQ. Ding EJ. "Stabilization Of The Less Stable Orbit By A Tiny Near-Resonance Periodic Signal" Physical Review A. 54(6):R5889-R5892, (1996)

[247] Vilar JMG. Perezmadrid A. Rubi JM. "Stochastic Resonance In A Dipole" Physical Review A. 54(6):6929-6932, (1996)

[248] A.Guderian, G.Dechert, K.-P.Zeyer and F.W.Schneider "Stochastic Resonance in Chemistry: The Belousov-Zhabotinsky Reaction" Journal of Physical Chemistry, 100(10), 4437-4441, (1996)

[249] A.Foerster, M.Merget and F.W.Schneider "Stochastic Resonance in Chemistry: The Peroxidase-Oxidase Reaction" Journal of Physical Chemistry, 100(10), 4442-4447, (1996)

[250] Apostolico F. Gammaitoni L. Marchesoni F. Santucci S. "Resonant Trapping - A Failure Mechanism In Switch Transitions" Physical Review E. 55(1 Part A):36-39, (1997)

[251] Longtin A. "Autonomous Stochastic Resonance In Bursting Neurons" Physical Review E. 55(1 Part B):868-876, (1997)

[252] Babinec P. "Stochastic Resonance In The Weidlich Model Of Public Opinion Formation" Physics Letters A. 225(1-3):179-181, (1997)

[253] Su C. Rangwala Aa. Wodkiewicz K. "Laser-Phase-Noise-Induced Stochastic-Resonance Fluorescence" Zeitschrift Fur Naturforschung Section A-A Journal Of Physical Sciences. 52 (1-2):127-129, (1997)

[254] Plesser He. Tanaka S. "Stochastic Resonance In A Model Neuron With Reset" Physics Letters A. 225(4-6):228-234, (1997)

[255] Pikovsky As. Kurths J. "Coherence Resonance In A Noise-Driven Excitable System" Physical Review Letters. 78(5):775-778, (1997)

[256] Simonotto E. Riani M. Seife C. Roberts M. Twitty J. Moss F. "Visual Perception Of Stochastic Resonance" Physical Review Letters. 78(6):1186-1189, (1997)

[257] Ricci Tf. Scherer C. "Linear Response And Stochastic Resonance Of Superparamagnets" Journal Of Statistical Physics. 86(3-4):803-819, (1997)

[258] Chapeaublondeau F. "Noise-Enhanced Capacity Via Stochastic Resonance In An Asymmetric Binary Channel" Physical Review E. 55(2):2016-2019, (1997)

[259] Chialvo Dr. Longtin A. Mullergerking J. "Stochastic Resonance In Models Of Neuronal Ensembles" Physical Review E. 55(2):1798-1808, (1997)

[260] Chapeaublondeau F. Godivier X. "Theory Of Stochastic Resonance In Signal Transmission By Static Nonlinear Systems" Physical Review E. 55(2):1478-1495, (1997)

[261] Bai M. Jeon D. Lee Sy. Ng Ky. Riabko A. Zhao X. "Stochastic Beam Dynamics In Quasi-Isochronons Storage Rings" Physical Review E. 55(3 Part B):3493-3506, (1997)

[262] Thorwart M. Jung P. "Dynamical Hysteresis In Bistable Quantum Systems" Physical Review Letters. 78(13):2503-2506, (1997)

[263] Bezrukov Sm. Vodyanoy I. "Stochastic Resonance In Non-Dynamical Systems Without Response Thresholds" (Vol 385, Pg 319, 1997) Nature. 386(6626):738, (1997)

[264] Vilar Jmg. Rubi Jm. "Spatiotemporal Stochastic Resonance In The Swift-Hohenberg Equation" Physical Review Letters. 78(15):2886-2889, (1997)

[265] Pareek Tp. Mahato Mc. Jayannavar Am. "Stochastic Resonance And Nonlinear Response In A Dissipative Quantum Two-State System" Physical Review B-Condensed Matter. 55(15) (1997)

[266] Foss J. Moss F. Milton J. Noise, "Multistability, And Delayed Recurrent Loops" Physical Review E. 55(4):4536-4543, (1997)

[267] Sides Sw. Ramos Ra. Rikvold Pa. Novotny Ma. "Kinetic Ising System In An Oscillating External Field - Stochastic Resonance And Residence-Time Distributions" Journal Of Applied Physics. 81(8 Part 2b):5597-5599, (1997)

[268] Inchiosa Me. Bulsara Ar. Gammaitoni L. "Higher-Order Resonant Behavior In Asymmetric Nonlinear Stochastic Systems" Physical Review E. 55(4):4049-4056, (1997)

[269] Castelpoggi F. Wio Hs. "Stochastic Resonance In Extended Systems - Enhancement Due To Coupling In A Reaction-Diffusion Model" Europhysics Letters. 38(2):91-95, (1997)

[270] Astumian Rd. "Thermodynamics And Kinetics Of A Brownian Motor" Science. 276(5314):917-922, (1997)

[271] Grifoni M., Hartmann L., Hanggi P. "Dissipative Tunneling With Periodic Polychromatic Driving - Exact Results And Tractable Approximations" Chemical Physics. 217(2-3), (1997)

[272] Mahato Mc., Jayannavar Am. "Relation Between Stochastic Resonance And Synchronization Of Passages In A Double-Well System" Physical Review E. 55(5 Part B):6266-6269, (1997)

[273] Grigorenko An., Nikitin Pi., Roschchepkin Gv. "Observation Of Stochastic Resonance In A Monostable Magnetic System Jetp" Letters. 65(10):828-832, (1997)

[274] Hartmann L., Grifoni M., Hanggi P. "Dissipative Transport In Dc-Ac-Driven Tight-Binding Lattices" Europhysics Letters. 38(7):497-502, (1997)

[275] Wang W., Wang Zd. "Internal-Noise-Enhanced Signal Transduction In Neuronal Systems" Physical Review E. 55(6 Part B):7379-7384, (1997)

[276] Porra Jm. When "Coherent Stochastic Resonance Appears" Physical Review E. 55(6 Part A):6533-6539, (1997)

[277] Eichwald C., Walleczek J. "Aperiodic Stochastic Resonance With Chaotic Input Signals In Excitable Systems" Physical Review E. 55(6 Part A):R6315-R6318, (1997)

[278] Marchesoni F. "Comment On Stochastic Resonance In Washboard Potentials" Physics Letters A. 231(1-2):61-64, (1997)

[279] Lindner Jf., Prusha Bs., Clay Ke. "Optimal Disorders For Taming Spatiotemporal Chaos" Physics Letters A. 231(3-4):164-172, (1997)

[280] Cabrera Jl., Delarubia Fj. "Resonance-Like Phenomena Induced By Exponentially Correlated Parametric Noise" Europhysics Letters. 39(2):123-128, (1997)

[281] Bose D., Sarkar Sk. "Noisy Bistable Hysteresis With Modulation Of Large Amplitude And High Frequency" Physics Letters A. 232(1-2):49-54, (1997)

[282] Chapeaublondeau F. "Input-Output Gains For Signal In Noise In Stochastic Resonance" Physics Letters A. 232(1-2):41-48, (1997)

[283] Collins Jj., Imhoff Tt., Grigg P. "Noise-Mediated Enhancements And Decrements In Human Tactile Sensation" Physical Review E. 56(1 Part B):923-926, (1997)

[284] Salman H., Braun E. "Voltage Dynamics Of Single-Type Voltage-Gated Ion-Channel Protein Ensembles" Physical Review E. 56(1 Part B):852-864, (1997)

[285] Grifoni M., Winterstetter M., Weiss U. "Coherences And Populations In The Driven Damped Two-State System" Physical Review E. 56(1 Part A):334-345, (1997)

[286] Neiman A., Saparin Pi., Stone L. "Coherence Resonance At Noisy Precursors Of Bifurcations In Nonlinear Dynamical Systems" Physical Review E. 56(1 Part A):270-273, (1997)

[287] Vilar Jmg., Rubi Jm. "Effect Of The Output Of The System In Signal Detection" Physical Review E. 56(1 Part A):R 32-R 35, (1997)

[288] Neiman A., Schimanskygeier L., Moss "F. Linear Response Theory Applied To Stochastic Resonance In Models Of Ensembles Of Oscillators" Physical Review E. 56(1 Part A):R 9-R 12, (1997)

[289] Fulinski A. "Stochastic Resonances In Active Transport In Biological Membranes" Acta Physica Polonica B. 28(8):1811-1825, (1997)

[290] Castro R., Sauer T. "Chaotic Stochastic Resonance - Noise-Enhanced Reconstruction Of Attractors" Physical Review Letters. 79(6):1030-1033, (1997)

[291] Astumian Rd., Adair Rk., Weaver Jc. "Stochastic Resonance At The Single-Cell Level" Nature. 388(6643):632-633, (1997)

[292] Bezrukov Sm. Vodyanoy I. "Stochastic Resonance At The Single-Cell Level – Reply" Nature. 388(6643):633, (1997)

[293] Mahato Mc., Jayannavar Am. "Two-Well System Under Large Amplitude Periodic Forcing - Stochastic Synchronization, Stochastic Resonance And Stability" Modern Physics Letters B. 11(19):815-820, (1997)

[294] Grigorenko An., Nikitin Pi., Roshchepkin Gv. "Frequency Mixing In A Bistable System In The Presence Of Noise" Journal Of Experimental & Theoretical Physics. 85(2):343-350, (1997)

[295] Torres Jl., Trainor L. "Stochastic Resonance And Computation" Journal Of Applied Physics. 82(5):2702-2703, (1997)

[296] Gitterman M., Shrager Ri., Weiss Gh. "Stochastic Resonance And Symmetry Breaking In A One-Dimensional System" Physical Review E. 56(3 Part B):3713-3716, (1997)

[297] Gade Pm., Rai R., Singh H. "Stochastic Resonance In Maps And Coupled Map Lattices" Physical Review E. 56(3 Part A):2518-2526, (1997)

[298] Yang Hl., Huang Zq., Ding Ej. "Noise, Order, And Spatiotemporal Intermittency" Physical Review E. 56(3 Part A):R2355-R2358, (1997)

[299] Sterken C., Degroot M., Vangenderen Am. "Cyclicities In The Light Variations Of Lbvs .1. The Multi-Periodic Behavior Of The Lbv Candidate Zeta(1)" Sco Astronomy & Astrophysics. 326(2):640-646, (1997)

[300] Grigorenko An. Nikitin Si. Roschepkin Gv. "Stochastic Resonance At Higher Harmonics In Monostable Systems" Physical Review A. 56(5 Part A):R4907-R4910, (1997)

[301] Casado Jm. "Noise-Induced Coherence In An Excitable System" Physics Letters A. 235(5):489-492, (1997)

[302] Grifoni M. Hartmann L. Berchtold S. Hanggi P. "Quantum Tunneling And Stochastic Resonance" (Vol 53, Pg 5890, 1996) Physical Review A. 56(5 Part B):6213, (1997)

[303] Hohmann W. Lebender D. Muller J. Schinor N. Schneider Fw. "Enhancement Of The Production Rate In Chemical Reactions With Thresholds" Journal Of Physical Chemistry. 101(48):9132-9136, (1997)

[304] Gailey Pc. Neiman A. Collins Jj. Moss F. "Stochastic Resonance In Ensembles Of Nondynamical Elements - The Role Of Internal Noise" Physical Review Letters. 79(23) (1997)

[305] Fulinski A. "Active Transport In Biological Membranes And Stochastic Resonances" Physical Review Letters. 79(24):4926-4929, (1997)

[306] Krawiecki A. "Stochastic Resonance In System With On-Off Intermittency" Acta Physica Polonica A. 92(6):1101-1108, (1997)

[307] Plata J. "Approximate Analytic Solutions For Noise-Induced Coherent Oscillations In Autonomous Nonlinear Systems" Pysical Review A. 56(6):6516-6523, (1997)

[308] Raikher Yl. Stepanov Vi. Grigorenko An. Nikitin Pi. "Nonlinear Magnetic Stochastic Resonance - Noise-Strength-Constant-Force Diagrams" Physical Review A. 56(6) (1997)

[309] Berdichevsky V. Gitterman M. "Josephson Junction With Noise" Physical Review A. 56(6):6340-6354, (1997)

[310] Yoden S. "Classification Of Simple Low-Order Models In Geophysical Fluid Dynamics And Climate Dynamics" Nonlinear Analysis-Theory Methods & Applications. 30(7):4607-4618, (1997)

[311] Zheng Zg. Hu Bb. Hu G. "Spatiotemporal Dynamics Of Discrete Sine-Gordon Lattices With Sinusoidal Couplings" Physical Review A. 57(1):1139-1144, (1998)

[312] Kaufman Ik. Luchinsky Dg. Mcclintock Pve. Soskin Sm. Stein Nd. "Zero-Dispersion Stochastic Resonance In A Model For A Superconducting Quantum Interference Device" Physical Review A. 57(1):78-87, (1998)

[313] Saltzman B. Hu Hj. Oglesby Rj. "Transitivity Properties Of Surface Temperature And Ice Cover In The Ccm1" Dynamics Of Atmospheres & Oceans. 27(1-4):619-629, (1998)

[314] Liu Hs. Chao Bf. "Wavelet Spectral Analysis Of The Earths Orbital Variations And Paleoclimatic Cycles" Journal Of The Atmospheric Sciences. 55(2):227-236, (1998)

[315] Lawrence Jk. Ruzmaikin Aa. "Transient Solar Influence On Terrestrial Temperature Fluctuations" Geophysical Research Letters. 25(2):159-162, (1998)

[316] Taft G. Makri N. "Effects Of Periodic Driving On Asymmetric Two-Level Systems Coupled To Dissipative Environments" Journal Of Physics B-Atomic Molecular & Optical Physics. 31(2):209-226, (1998)

[317] Loreto V. Serva M. Vulpiani A. "On The Concept Of Complexity Of Random Dynamical Systems" International Journal Of Modern Physics B. 12(3):225-243, (1998)

[318] Gourier D. Gerbault D. Stochastic Resonance In An Electron-Spin-Nuclear-Spin System Physical Review B-Condensed Matter. 57(5):2679-2682, (1998)

[319] Sarkar P. Bhattacharyya Sp. "Noise Induced Tunneling Suppression Of A Metastable Quantum System Driven By An Oscillating Field - A Numerical Experiment" Physics Letters A. 238(2-3):141-146, (1998)

[320] Berdichevsky V. Gitterman M. "Stochastic Resonance And Ratchets - New Manifestations" Physica A. 249(1-4):88-95, (1998)

[321] Kadar S. Wang Jc. Showalter K. "Noise-Supported Travelling Waves In Sub-Excitable Media" Nature. 391(6669):770-772, (1998)

[322] Moss F. "Chemical Dynamics - Noisy Waves" Nature. 391(6669):743-744, (1998)

[323] Vanwiggeren Gd. Roy R. "Communication With Chaotic Lasers" Science. 279, (1998)

[324] Inchiosa Me. Bulsara Ar. Hibbs Ad. Whitecotton Br. "Signal Enhancement In A Nonlinear Transfer Characteristic" Physical Review Letters. 80(7):1381-1384, (1998)

[325] Gonzalez Ja. Mello Ba. Reyes Li. Guerrero Le. "Resonance Phenomena Of A Solitonlike Extended Object In A Bistable Potential" Physical Review Letters. 80(7):1361-1364, (1998)

[326] Vigreux C. Binet L. Gourier D. "Bistable Conduction Electron Spin Resonance In Metallic Lithium Particles" Journal Of Physical Chemistry B. 102(7):1176-1181, (1998)

[327] Zhang Y. Hu G. Liu H. Xiao Jh. "Collective Behavior In Globally Coupled Systems Consisting Of Two Kinds Of Competing Cells" Physical Review A. 57(3 Part A):2543-2548, (1998)

[328] Wang W. Wang Yq. Wang Zd. "Firing And Signal Transduction Associated With An Intrinsic Oscillation In Neuronal Systems" Physical Review A. 57(3 Part A):R2527-R2530, (1998)

[329] Litong M. Saloma C. "Detection Of Subthreshold Oscillations In A Sinusoid-Crossing Sampling" Physical Review A. 57(3 Part B):3579-3588, (1998)

[330] Lee Sg. Neiman A. Kim S. "Coherence Resonance In A Hodgkin-Huxley Neuron" Physical Review A. 57(3 Part B):3292-3297, (1998)

[331] Sadykov Ek. Isavnin Ag. Boldenkov Ab." On The Theory Of Quantum Stochastic Resonance In Single-Domain Magnetic Particles" Physics Of The Solid State. 40(3):474-476, (1998)

[332] Liu F. Wang W. "Stochastic Resonance In A Coupled Neuronal Network" Chinese Physics Letters. 15(2):152-154, (1998)

[333] Alarcon T. Perezmadrid A. Rubi Jm. "Periodic Modulation Induced Increase Of Reaction Rates In Autocatalytic Systems" Journal Of Chemical Physics. 108(17):7367-7374, (1998)

[334] Vilar Jmg. Sole Rv. "Effects Of Noise In Symmetric Two-Species Competition" Physical Review Letters. 80(18):4099-4102, (1998)

[335] Buchleitner A. Mantegna Rn. "Quantum Stochastic Resonance In A Micromaser" Physical Review Letters. 80(18):3932-3935, (1998)

[336] Nazario Ma. Saloma C. "Signal Recovery In Sinusoid-Crossing Sampling By Use Of The Minimum-Negativity Constraint" Applied Optics. 37(14):2953-2963, (1998)

[337] Bezak V. "An Exactly Solvable Linear Model Giving Indication Of Stochastic Resonance" Czechoslovak Journal Of Physics. 48(5):529-535, (1998)

[338] Sung Wy. Park Pj. "Transition Dynamics Of Biological Systems On Mesoscopic Scales - Effects Of Flexibility And Fluctuations" Physica A. 254(1-2):62-72, (1998)

[339] Dykman Mi. Maloney Cm. Smelyanskiy Vn. Silverstein M. "Fluctuational Phase-Flip Transitions In Parametrically Driven Oscillators" Physical Review A. 57(5 Part A), (1998)

[340] Kuperman Mn. Wio Hs. Izus G. Deza R. "Stochastic Resonant Media - Signal-To-Noise Ratio For The Activator-Inhibitor System Through A Quasivariational Approach" Physical Review A. 57(5 Part A):5122-5125, (1998)

[341] Castelpoggi F. Wio Hs. "Stochastic Resonant Media - Effect Of Local And Nonlocal Coupling In Reaction-Diffusion Models" Physical Review A. 57(5 Part A):5112-5121, (1998)

[342] Alarcon T. Perezmadrid A. Rubi Jm. "Stochastic Resonance In Nonpotential Systems" Physical Review A. 57(5 Part A):4979-4985, (1998)

[343] B. Andò, S. Baglio, S. Graziani, N. Pitrone, Threshold error reduction in linear measurement devices by adding noise signal, IMTC'98, St. Paul Minnesota, USA May 18-21, (1998)

[344] B. Andò, S. Baglio, S. Graziani, N. Pitrone, Threshold error reduction in optical displacement sensors, via Stochastic resonance, IMEKO '98, Napoli, (1998).

[345] Locher M. Cigna D. Hunt Er. "Noise Sustained Propagation Of A Signal In Coupled Bistable Electronic Elements" Physical Review Letters. 80(23):5212-5215, (1998)

[346] Deco G. Schurmann B. "Stochastic Resonance In The Mutual Information Between Input And Output Spike Trains Of Noisy Central Neurons" Physica D. 117(1-4):276-282, (1998)

[347] Krawiecki A. "Stochastic Resonance In On-Off Intermittency" Acta Physica Polonica B. 29(6):1589-1598, (1998)

[348] Shillcock J. Seifert U. "Escape From A Metastable Well Under A Time-Ramped Force" Physical Review A. 57(6):7301-7304, (1998)

[349] Fakir R. "Nonstationary Stochastic Resonance" Physical Review A. 57(6):6996-7001, (1998)

[350] Barzykin Av. Seki K. Shibata F. "Periodically Driven Linear System With Multiplicative Colored Noise" Physical Review A. 57(6):6555-6563, (1998)

[351] Sides Sw. Rikvold Pa. Novotny Ma." Stochastic Hysteresis And Resonance In A Kinetic Ising System" Physical Review A. 57(6):6512-6533, (1998)

[352] Galdi V. Pierro V. Pinto Im. "Evaluation Of Stochastic-Resonance-Based Detectors Of Weak Harmonic Signals In Additive White Gaussian Noise" [Review] Physical Review A. 57(6), (1998)

[353] Choi Mh. Fox Rf. Jung P. "Quantifying Stochastic Resonance In Bistable Systems - Response Vs Residence-Time Distribution Functions" Physical Review A. 57(6):6335-6344, (1998)

[354] Kim Yw. Sung W. "Does Stochastic Resonance Occur In Periodic Potentials" Physical Review A. 57(6):R6237-R6240, (1998).

[355] Amemiya T. Ohmori T. Nakaiwa M. Yamaguchi T. "Two-Parameter Stochastic Resonance In A Model Of The Photosensitive Belousov-Zhabotinsky Reaction In A Flow System" Journal Of Physical Chemistry. 102(24):4537-4542, (1998)

[356] Gong Yf. Xu Jx. Hu Sj. "Stochastic Resonance - When Does It Not Occur In Neuronal Models" Physics Letters A. 243(5-6):351-359, (1998)

[357] Nozaki D. Yamamoto Y. "Enhancement Of Stochastic Resonance In A Fitzhugh-Nagumo Neuronal Model Driven By Colored Noise" Physics Letters A. 243(5-6):281-287, (1998)

[358] Vilar Jmg. Gomila G. Rubi Jm. "Stochastic Resonance In Noisy Nondynamical Systems" Physical Review Letters. 81(1):14-17, (1998)

[359] Gourier D. Binet L. Gerbault D. "Hysteresis And Memory Of The Magnetic Resonance Of Conduction Electrons In Solids - From Bistability To Stochastic Resonance" Applied Magnetic Resonance. 14(2-3):183-201, (1998)

[360] Goychuk I. Grifoni M. Hanggi P. "Nonadiabatic Quantum Brownian Rectifiers" Physical Review Letters. 81(3):649-652, (1998)

[361] Mato G. "Stochastic Resonance In Neural Systems - Effect Of Temporal Correlation In The Spike Trains" Physical Review A. 58(1):876-880, (1998)

[362] Acharyya M. "Nonequilibrium Phase Transition In The Kinetic Ising Model - Is The Transition Point The Maximum Lossy Point" Physical Review A. 58(1):179-186, (1998)

[363] Inchiosa Me. Bulsara Ar. "Dc Signal Detection Via Dynamical Asymmetry In A Nonlinear Device" Physical Review A. 58(1):115-127, (1998)

[364] Izus Gg. Deza Rr. Wio Hs. "Exact Nonequilibrium Potential For The Fitzhugh-Nagumo Model In The Excitable And Bistable Regimes" Physical Review A. 58(1):93-98, (1998)

APPENDIX B

APPENDIX B
THE χ^2 TEST

The χ^2 test is an analytical method of statistical investigation with which it is possible to establish the distribution probability with which a series $\{x_k$ of N data can be approximated with a known distribution [1].
It is essentially based on calculation of the discrete function

$$q \equiv \sum_{i=1}^{n} \frac{(n_0^i - n_e^i)^2}{n_e^i} \tag{B.1}$$

where:
n is the total number of groups into which the data set is divided;
n_0^i is the number of data items belonging to the *i-th* group;
n_e^i is the number of data items that would be observed in the *i-th* group if the distribution were perfectly *Gaussian*;

For greater clarity, it is necessary to analyse how n_e^i is calculated.
Using one of the tables to be found in the literature, e.g. the table shown in Figure B.1, and knowing the number of data items, N, to which the test is to be applied, it is possible to determine the number of groups into which they have to be divided.
Observe that if N>40 it is possible to apply the following analytical expression [1]

$$n=1.87(N-1)^{0.4} \tag{B.2}$$

With the aid of comparison tables, calculation of the q function determines the minimum index value, for the vector of points being examined, to assure a certain degree of confidence that the distribution is *Gaussian*, given the degree of freedom of the data.
We define the following probability

$F(w)$= the probability that a data item will be in the range]$-\infty, w$]

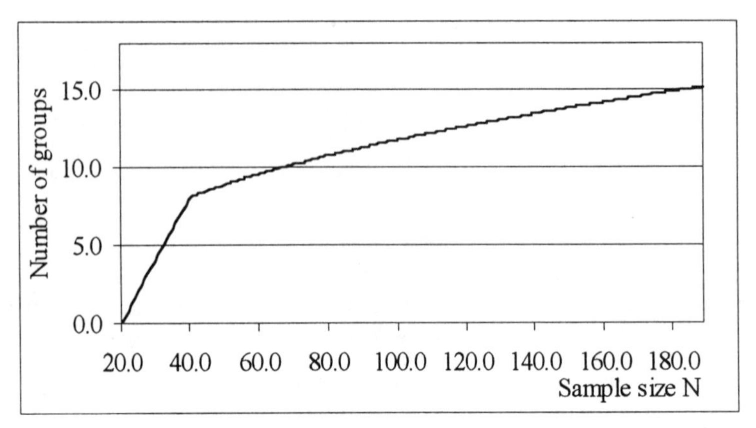

Figure B.1. Knowing the number of data items, N, to which the test is to be applied, it is possible to determine the number of groups into which they have to be divided.

where

$$w_k = \frac{x_k - \mu}{\sigma}, \text{ with } k=1, \ldots, N \tag{B.3}$$

and x is a generic element of the distribution of points being considered, μ is the mean value and σ the standard deviation. With a change in variable (B.3) the data sequence is normalised to one with a null mean and variance of 1. At this point, n_e^i is calculated on the basis of the following expression

$$n_e^i = N * F(w_j) \text{ with } w_j \text{ elements belonging to the i-th group}$$

Let us assume, for example, that we have a sequence of 20 data items with a mean μ of 10.11 and a standard deviation σ of 0.14, the distribution of which is Gaussian. We wish to calculate n_e for the first group of values, which we assume contains the elements $x \in \left]-\infty, 10.03\right]$. Using relation (B.3) we get the corresponding range of values

$w \in \left]-\infty, -0.572\right]$. The probability *F(w)* that a value falls into the range $\left]-\infty, -0.572\right]$ coincides with the probability that it falls into $\left]-\infty, 10.03\right]$.
From *F(w)* curve, shown in Figure B.2, which is valid for *Gaussian* distributions, we find that when w=-0.572 the probability $P(-\infty < w < 0.572)$ is 0.717.
We therefore have

$P(-\infty<w<-0.572)=P(0.572<w<\infty)=1-P(-\infty<w<0.572)=0.283$, hence

n_e^i =(20*0.283)=5.66= number of elements in the data sequence that should fall into the range $\left]-\infty, 10.03\right]$.

Once the quantities appearing in (B.1) are known, it is possible to calculate the value taken by q for the given distribution. Obviously, calculation of the index q varies according to the type of distribution.
To complete the test it is necessary to establish a confidence range within which it is possible to consider valid the hypothesis that the distribution is the assumed one and calculate the degrees of freedom from the following relation

$$degrees\ of\ freedom = number\ of\ groups - 3.$$

With this information we can use tables available in the literature, e.g. Table B.1 [2], or analytical expressions to obtain the quantity $\chi^2_{1-\alpha}(n)$ with which the value obtained for q can be compared.

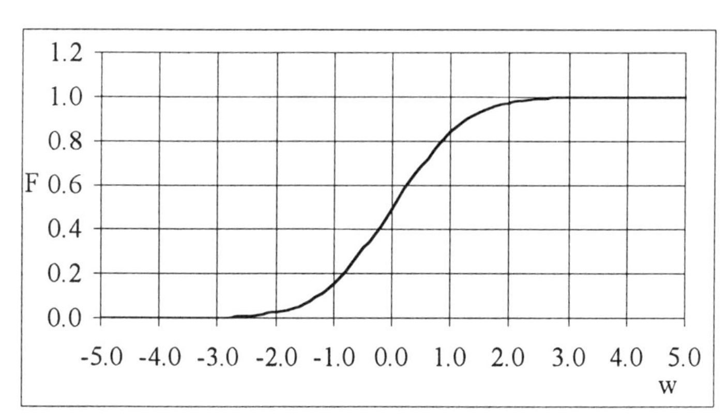

Figure B.2. The probability F(W) in case of a *Gaussian* distribution.

More specifically, to calculate the quantity $\chi^2_{1-\alpha}(n)$. We can use the following expression [1]

$$\chi^2_{1-\alpha}(n) = 0.5 * (Z_u + \sqrt{2n-1})^2 \qquad \text{(B.4)}$$

denoting by Z_u the u percentile of the standard normal density.
If $q \leq \chi^2_{1-\alpha}$ it is possible to sate that with the confidence range established the distribution of the data sequence is the same as the one hypothesised.

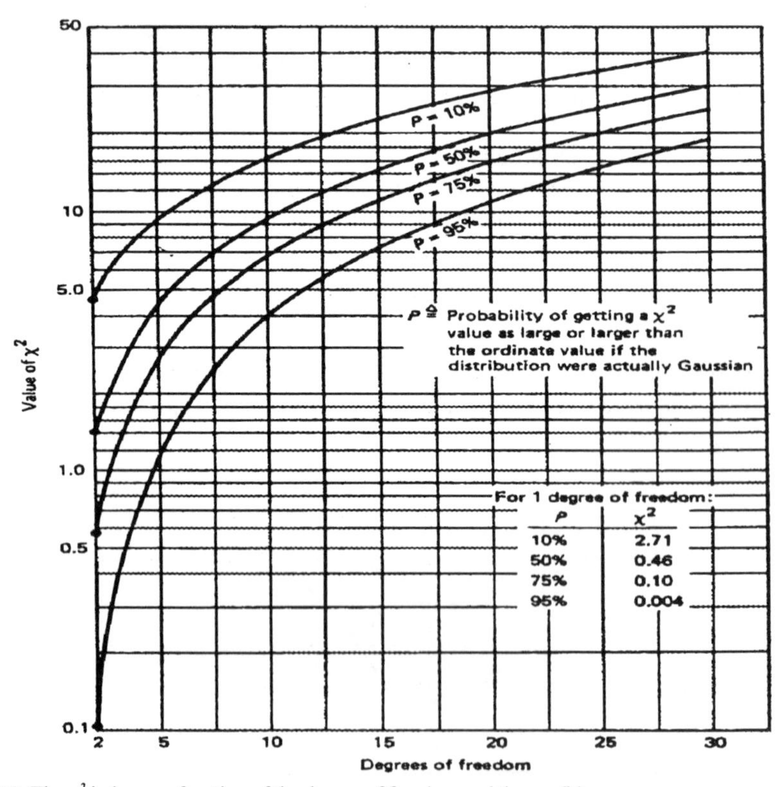

Table B.1. The χ^2 index as a function of the degree of freedom and the confidence range.

References

[1] A. Papoulis, *"Probability, Random Variables, and Stochastic Process"*, McGRAW-HILL BOOK COMPANY

[2] E. O. Doebelin, *"Measurement Systems"*, McGRAW-HILL BOOK COMPANY, third edition, 1985.

REFERENCES BY KEYWORDS